Georg Neumann

Nutrition in Sport

Meyer & Meyer Sport

Original title: Ernährung im Sport
2. überarbeitete Auflage
– Aachen: Meyer und Meyer Verlag, 1998
Translated by Paul D. Chilvers-Grierson

British Library Cataloguing in Publication Data
A catalogue for this book is available from the British Library

Neumann, Georg :
Nutrition in Sport/ Georg Neumann.
[Transl.: Paul D. Chilvers-Grierson].
– Oxford : Meyer & Meyer Sport (UK) Ltd., 2001
ISBN 1-84126-003-7

Oxford, Aachen, Olten (CH), Vienna, Québec,
Lansing/Michigan, Adelaide, Auckland, Johannesburg, Budapest
Member of the World
Sportpublishers' Association

Cover Photo: Sportpressefoto Bongarts, Hamburg
Photos & Illustrations: Georg Neumann, Leipzig
Cover design: Walter Neumann, N&N Design-Studio, Aachen
Cover and type exposure: frw, Reiner Wahlen, Aachen
Editorial: John Coghlan, Jürgen Schiffer
Printed and bound in Germany
by Burgverlag Gastinger GmbH, Stolberg
ISBN 1-84126-003-7
e-mail: verlag@meyer-meyer-sports.com

1 Introduction

To maintain the ability to stay alive constant ingestion of food is necessary. Although most people in Europe eat normally, new forms of **poor nutrition** are increasing. The risks of poor nutrition include excess weight, increased blood pressure and increased concentration of blood fats, among others. The term poor nutrition can mean too much food ingestion, but also a deficit of energy providers (e.g. proteins) or a deficiency of active substances (e.g. vitamins, minerals).

In the industrial countries nutrition is a central problem because, in conjunction with the lack of exercise, it is a key contributor to excess weight and thus becomes a **risk factor** for many people. A sensible alternative is sport. With increasing physical activity, changes in diet are necessary, about which there is a major information deficit.

Ingested foods contain nutrients. The **main** or **basic nutrients** include carbohydrates, proteins and fats. They create the operative foundation for increased muscular performance capacity in sport. Sporting activity is always linked with increased **energy utilisation** and accelerated **restructuring** in the muscles. Because nutrition is of central importance for sport participants, numerous authors have written about the subject. The basis for these writings are findings in epidemiology, nutritional physiology, practical experience and the effects of poor nutrition. Not all dietary habits have had their worth or lack of it scientifically verified; experience dominates.

One of the known facts is that people's eating habits only change very slowly. Sport practitioners also maintain their individual **dietary rituals**, especially if by chance they were linked with sporting successes. Apart from that, however, interest and willingness to assimilate new information and findings on nutritional physiology and to experiment on a personel level are increasing. Interested people include sport practitioners, coaches and trainers. The available knowledge and recommendations are mainly directed towards these people.

Applying new findings in sports nutrition in no way means greatly restricting **dietary habits** with regard to taste, aroma, appearance and processing of nutrients. For sport practitioners it is a matter of selecting at the appropriate time the most appropriate nutrients with regard to quantity and quality. This may not always coincide with personal appetite, but it can be useful for doing the particular sport and ensuring regeneration.

Nutrition is greatly influenced by beliefs, philosophy, myth, extreme variants and other factors. Sports nutrition, too, is not completely untouched by these influences. The recommendation derived from physiological findings to ingest greater proportions of carbohydrates in athletes' nutrition has led to alternatives being tried out. In this way **alternative dietary forms** have also gained a foothold in competitive sport. These dietary forms include vegetarianism, whole food nutrition, wholemeal foods, anthroposophic nutrition, macrobiotics, the performance diet (Dr Haas) among others. The meatless or low meat dietary forms mentioned here have a high proportion of carbohydrates. For sport practitioners, however, they are not without risk. The problem is in the **deficiency** of certain minerals (e.g. iron, calcium, iodine) and qualitatively valuable amino acids. If the pros and cons of these alternative dietary forms are objectively analysed and accepted, then nutritional physiological findings show that deliberate **supplementation** with active substances is absolutely necessary. The additional ingestion of the missing substances allows these varying dietary forms to be applied in conjunction with competive training as well.

Meanwhile, sports nutrition can call on comprehensive findings and experience gathered over the last ten years. It should be mentioned, though, that certain nutrional science findings have been taken out of their context or represented in great detail as single facts. Classic examples of this are **cholesterol** or the **saturated fatty acids** in nutrition. On the basis of findings mainly gathered from animal experiments, excessive ingestion of cholesterol and saturated fatty acids can encourage the development of coronary heart disease. The level of blood cholesterol is considered an indicator of the state of fat metabolism. Values under 200 mg/dl are called for as normal (ASSMANN, 1991). If this were rigidly put into practice it would mean that over 80% of all middle aged and older people would have pathological cholesterol values and be on medication (BERGER, 1994). Even more drastic is APFELBAUM (1994) who says people without congenital fat metabolism disruptions should not bother about their cholesterol level.

To clarify the great differences in opinion on cholesterol it should be mentioned that in healthy people the ingestion of cholesterol via food is 87% self-regulated by the liver. It is not intended to discuss here the complexity of the possible causes of hardening of the arteries; in no way is cholesterol alone responsible for it. For **cholesterol** sports training has the advantage that in the form of endurance training it lowers high fat metabolism values. Longer endurance training increases the activity of the enzymes that break down fat

(MARNIEMI et al., 1980, LITHELL et al., 1981). The concentration of HDL (high density lipoprotein) cholesterol which supports the transport of cholesterol back to the liver is highest in marathon runners (ADNER/ CASTELLI, 1980). Only endurance athletes, however, show a positive effect on fat metabolism, not athletes without a major endurance component in training load.

In sport the main purpose of nutrition is to ensure enough energy for load and regeneration after it. The essence of sports nutrition consists not only in the recommendation of **nutrient ratios**, i.e. of the energy proportions of carbohydrates, proteins and fats. A major aspect is ensuring sufficient energy to meet the demands of load and thus sufficient energy supply. An additional factor is the **nutrient density** of food ingested. Nutrient density refers to the proportions of vitamins, minerals (trace elements), fibre as well as the taste and appearance of food. Optimal athlete nutrition means there is constant consideration of the qualitative possibilities for the supply of energy and active substances, especially from the point of view of the sport being practised. The orientation yardstick is securing training load and not nutrition itself! Just like healthy untrained persons, sport practitioners must keep to the basic rules of physiologically sensible nutrition.

The common situations of iodine or calcium **deficiency** affect athletes just as much as untrained persons.

Chronic **calcium** deficiency hinders bone development and leads to the development of osteoporosis. In competitive sport young girls are especially affected, in whom hormonal disturbances also occur. In this connection it is necessary to know which foodstuffs are especially rich in **calcium** and how much must be ingested in relationship to the level of training load. The calcium requirement of 800 mg/day can of course be covered with 1 litre of milk.

It is the intention of this book to give interested sports practitioners, and those working with them in the broadest sense, ideas for dietary planning in sport. Through the summarising of experience it is meant to be both an orientation aid and a reference work. Optimal nutrition of competitive athletes includes the right to use supplementation with legal substances. Supporting one's health in a highly taxed state by ingesting additional substances is a preventive measure for competitive athletes designed to reduce risks. The industry has recognised this, and the diet market, and advertising for supplements, are correspondingly major factors.

In this book the basics of nutritional physiology will only be touched on briefly. At the same time, however, findings of nutritional science are the foundation for the description of sports nutrition and general orientation for nutrition. It is intentional that there are **no detailed recipes** for dietary composition; there are good standard books for this purpose.

Sports nutrition is not tied to any puritanical diet or lifestyle. Athletes can also consume luxury goods. During their active sporting career with competitive objectives, however, **alcohol consumption** should be nil. Adult athletes can still join in with the usual glass of victory champagne. Occasionally drinking low alcohol, or non-alcoholic, beer in the evening will also not damage performance capacity. Often a glass of beer serves as a sleeping aid. To avoid misundertandings: young people, on the other hand, should rigidly go without alcohol and nicotine during their sporting career.

"Fast food"is also popular amongst athletes. It should not be generally rejected on account of the lack of active substances (vitamins and minerals) and its high fat content (trans-fatty acids). But this one-sided, high-calorie diet should not be the main source of energy for sports practitioners. Young people in particular succumb to the temptations of all sorts of forms of poor nutrition; they usually have little time for eating and are not conscious of the necessity of healthy nutrition.

For competitive athletes, or people training in competitive sport, balancing the ingested nutrients, vitamins and minerals is always a personal matter. Generally speaking, the **objective of sports nutrition** is the optimisation of sporting potential, ensuring the processing of load (adaptation) and accelerated regeneration. Improved performance capacity in sport is only possible through training. Nutrition alone will not lead to any increase in performance capacity. No performance increase can be reached through excessive ingestion of vitamins, minerals or other active subtances. Optimal athlete's nutrition does not replace regular training.

2 Energy Metabolism

The performance capacity of the muscles depends on the constant **supply of energy**. The muscles draw their energy from carbohydrates, fats and protein ingested in food. If there is no supply of energy, a number of energy stores are available for muscle work. The energy store in the muscles are:

Adenosine triphosphate (ATP), creatine phosphate (CP), glycogen and trigylcerides (TG).

The working muscles are further supported by substrate exchange, especially by the **liver and fatty tissue**. During intensive short term load the liver converts its **glycogen** into glucose and thus secures the blood sugar concentration for a certain time. Blood sugar is a useful substrate for the exerted muscles and the brain.

The energy used in muscle work is mainly released as heat, and in a warm environment leads to sweating. Of the energy made use of, only 20 to 25% is available for muscle work, the other 75% is released as heat. The degree of energy exploitation is expressed in the degree of **effectiveness** (Eta) (Table 1). In sport-specific competitive training the degree of effectiveness of muscle work improves. In unspecific training, and in training with an emphasis on strength endurance, the degree of effectiveness of the muscle decreases.

During muscle contraction **energy rich ATP** is broken down into the energy poorer phosphate compound **adenosine diphosphate** (ADP). The energy released is used for muscle work. Several possibilities are available for the re-synthesis of ADP to ATP. The **substrates** providing energy for the working skeletal muscle are creatine phosphate, glucose, the free fatty acids and some amino acids (Fig. 1). Which substrates are used for the regeneration of ATP energy depends on the intensity and duration of muscle work. The creation of energy per time unit necessary for ATP re-synthesis is the key element for the use of the particular substrate (Fig. 2).

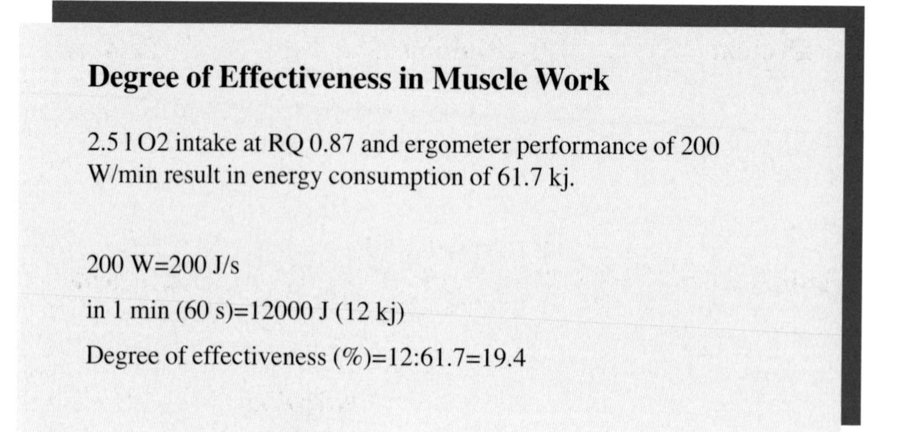

Degree of Effectiveness in Muscle Work

2.5 l O2 intake at RQ 0.87 and ergometer performance of 200 W/min result in energy consumption of 61.7 kj.

200 W=200 J/s

in 1 min (60 s)=12000 J (12 kj)

Degree of effectiveness (%)=12:61.7=19.4

Table 1

As the diagram in Fig. 2 shows, eleven times as much energy can be won in a minute from the combustion of glucose (4.4 mmol/min) as from fatty acids (0.4 mmol/min).

Fatty acids deliver the least energy per time unit and are inexhaustible during long duration load because of their large reserves (Fig. 3). On average athletes have fat reserves of 5-20 kg. To secure very long duration muscle work, constant **carbohydrate ingestion** is necessary, especially if the glycogen stores are exhausted.

2.1 Carbohydrates

Carbohydrates are of great significance for securing the performance capacity of the muscles. They are the most important **energy supplier** when their storage form, muscle and liver glycogen, is exhausted. Energy-wise, in the mitochondria only glucose is utilised directly, other carbohydrates must first be split up into glucose (see Fig. 3) The aerobic breakdown of glycogen to glucose is more effective than anaerobic breakdown. If during intensive load the muscle needs more energy for resynthesis of ATP, aerobic metabolism is replaced by anaerobic metabolism. Hereby the glucose is broken down to lactate (lactic acid). **Anaerobic glucose breakdown**, also called glycolysis, takes place without oxygen.

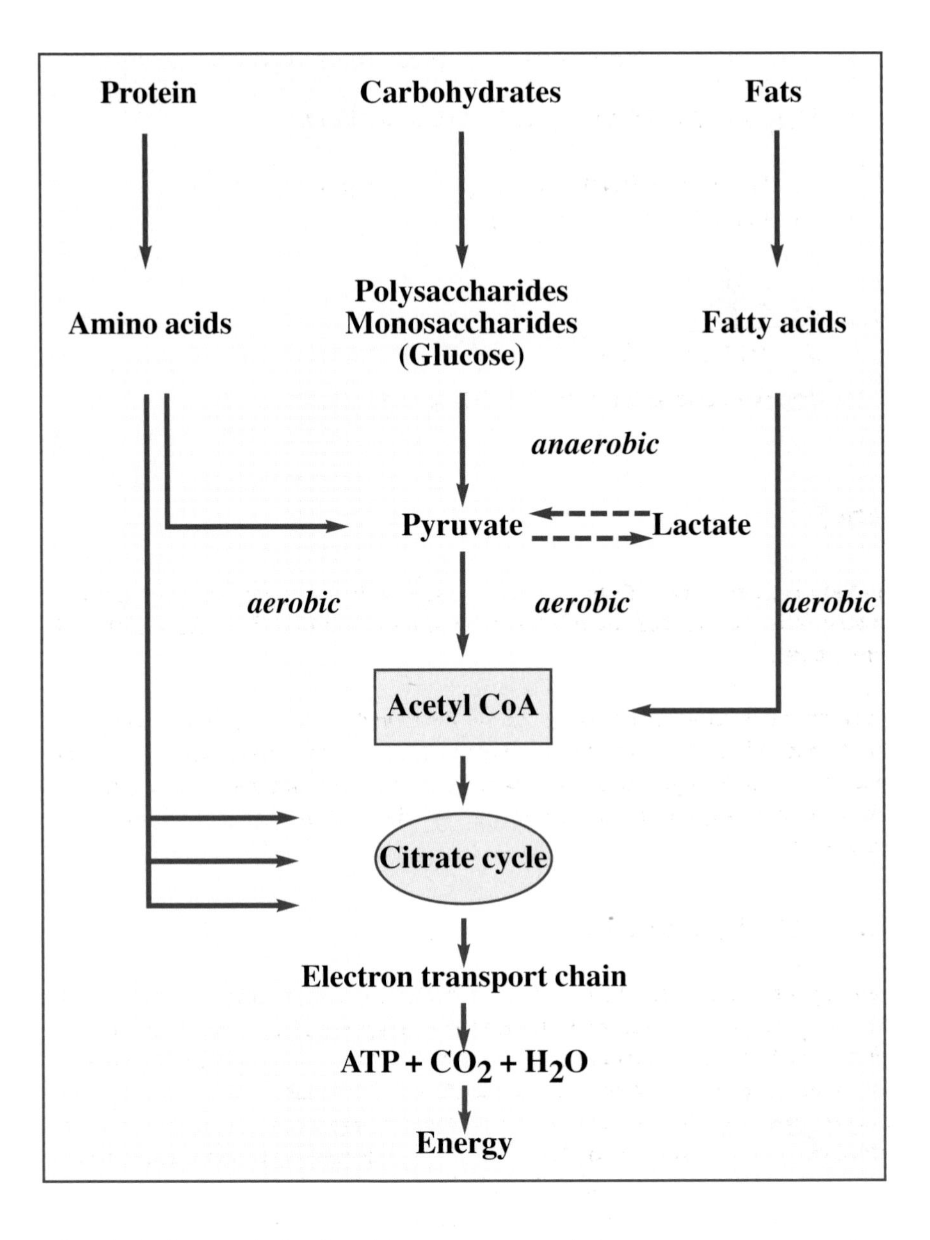

Fig. 1: Scheme of the breakdown patterns of the main energy suppliers carbohydrates, proteins and fats (free fatty acids)

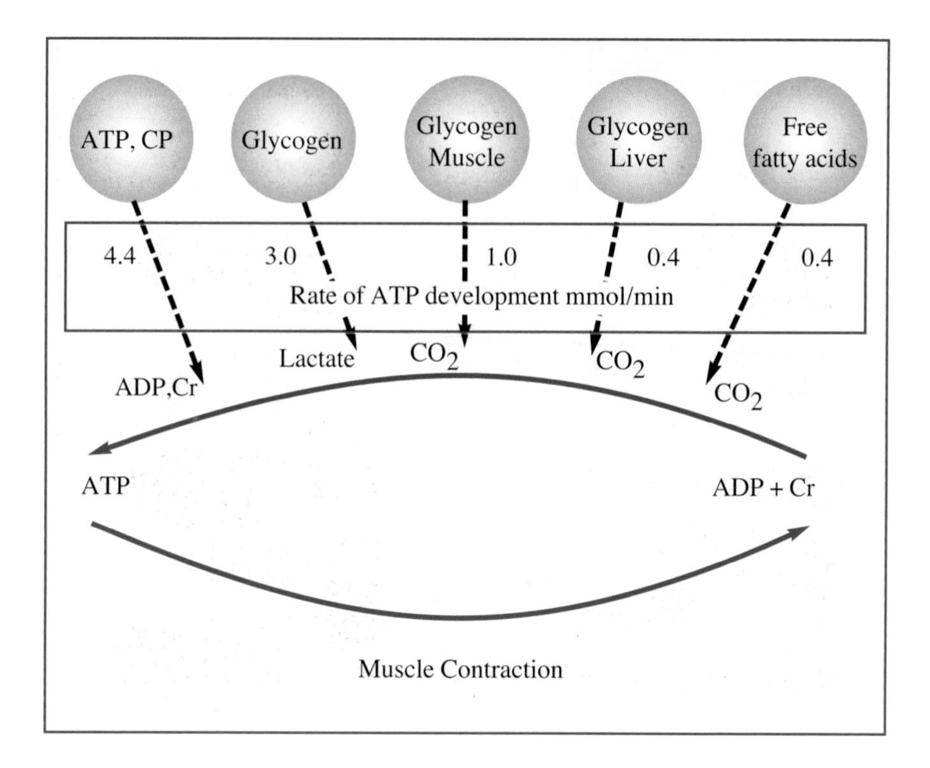

Fig. 2: Speed of energy conversion during ATP (adenosine triphosphate) resynthesis by the available energy suppliers. CP = creatine phosphate, ADP = adenosine diphosphate, Cr = creatine

As a result of regular training the glycogen stores in the loaded muscles and in the liver increase in size. The available glycogen stores in the muscles are about 250 g in untrained persons. Endurance training can increase them to about 400 g (see Fig. 3). The glycogen stores in the liver can also expand from about 80 g to 120 g through endurance training. With these glycogen stores of 500-520 g altogether, 2,050-2,130 kcal of energy could be won. The glycogen stores allow endurance athletes to achieve intensive loads of 90 to 120 minutes duration without eating. Loads spread over several hours call for additional energy ingestion in the form of carbohydrates (Fig. 4).

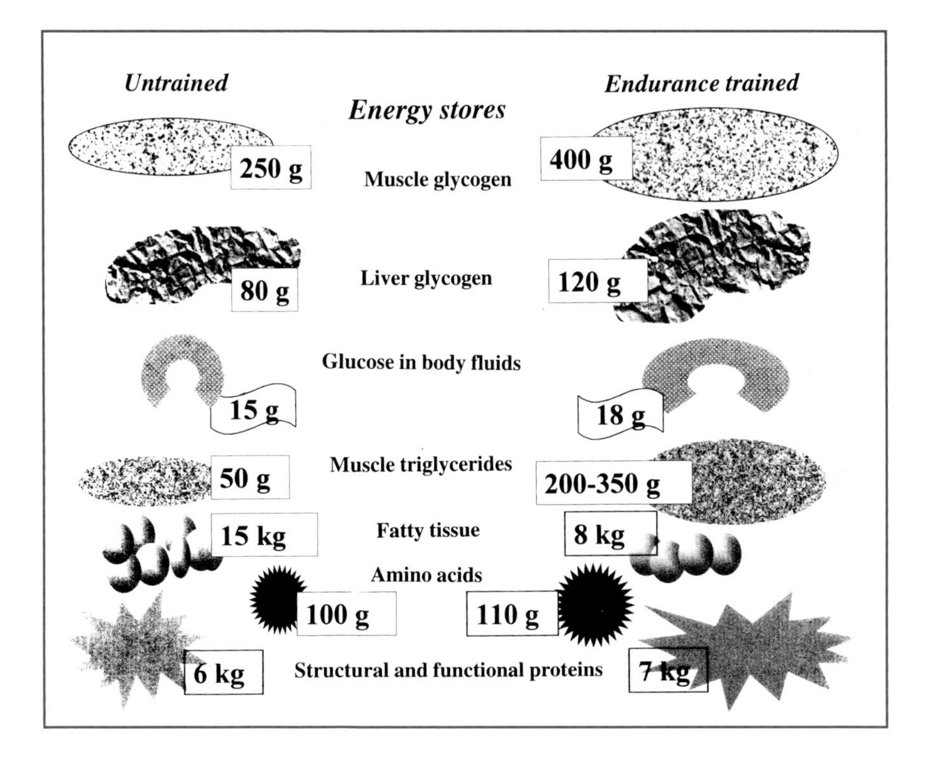

Fig. 3: Energy stores in the body. As a result of endurance training the energy stores of glycogen in the muscles and liver, as well as the triglyceride stores in the muscles, are enlarged. Endurance athletes' fatty tissue decreases and their muscle mass increases.

2.2. Fats

After carbohydrates, the breakdown products of fats, the free fatty acids are a major supplier of energy for the muscles at rest and under load. The **free fatty acids** (FFA) are practically inexhaustible energy suppliers for athletes.

The disadvantage of the fatty acids is that in comparison to glucose they lead to slow ATP re-synthesis. The **average fat reserves** of 6 to 15 kg in competitive athletes allow extreme long duration endurance loads (e.g. 100 km run, ultra triathlon, multiple ultra triathlon, 24 hour run etc.).

From a fat store of 10 kg an average of about 70,000 kcal could be released. If there was only muscular fat combustion, this energy would make it possible to run 23 marathons. Extreme sporting load is only possible with the help of trained fat metabolism.

For example, in the ultra triathlon (3.8 km swimming, 180 km cycling and marathon) of 8-12 hours duration, fatty acid combustion secures 65-75% of energy needs.

Fat metabolism can be trained. In competitive sport it calls for a training load of two hours and more. The glycogen stores must first be more or less exhausted by the longer load so that fatty acid combustion gains predominance, and the enzymes of fatty acid utilisation increase.

During longer load there is a shift in the proportions of energy production caused by a switch from mainly carbohydrate to mainly fat combustion (Fig. 4).

2.3. Proteins

Proteins belong to the third group of indispensable nutrients and are limited energy suppliers for muscle work. The **breakdown of proteins** to amino acids, however, only becomes effective in **stenuous emergencies**. The involvement of amino acids in energy metabolism takes place during long duration load lasting several hours.

Daily ingestion of a certain amount of protein is vital for life because the amino acid in the used up structures have to be replaced. Depending on the state of load of the muscles, 2-6% of the amino acids can be replaced daily. Thus the muscle structures exerted by training are restructured in about 50 days.

The German Association for Nutrition (Deutsche Gesellschaft für Ernährung/DGE) has made recommendations on the amount of proteins that should be ingested, namely 0.8 to 1.2 g/kg body mass for normal individuals. These recommendations are too low for competitive athletes. The **amount of protein** a person in training needs on a daily basis depends on the load, the type of sport and gender. On average it is 1.5 to 2.5 g/kg body mass (Fig. 5). Thus a male athlete with 80 kg body mass needs 120 to 200 g, and a female athlete with 55 kg needs an average of 83 to 138 g of protein per day of load.

At the finish athletes help themselves to water

Fast ladies at the start of the duathlon in St. Wendel

Exchanging experiences after finishing

Versatile sporting talent Sonja Krolik finishing

Dr Rainer Müller, European short triathlon champion in 1995, checks his training performance with a heart rate monitor and portable lactate set.

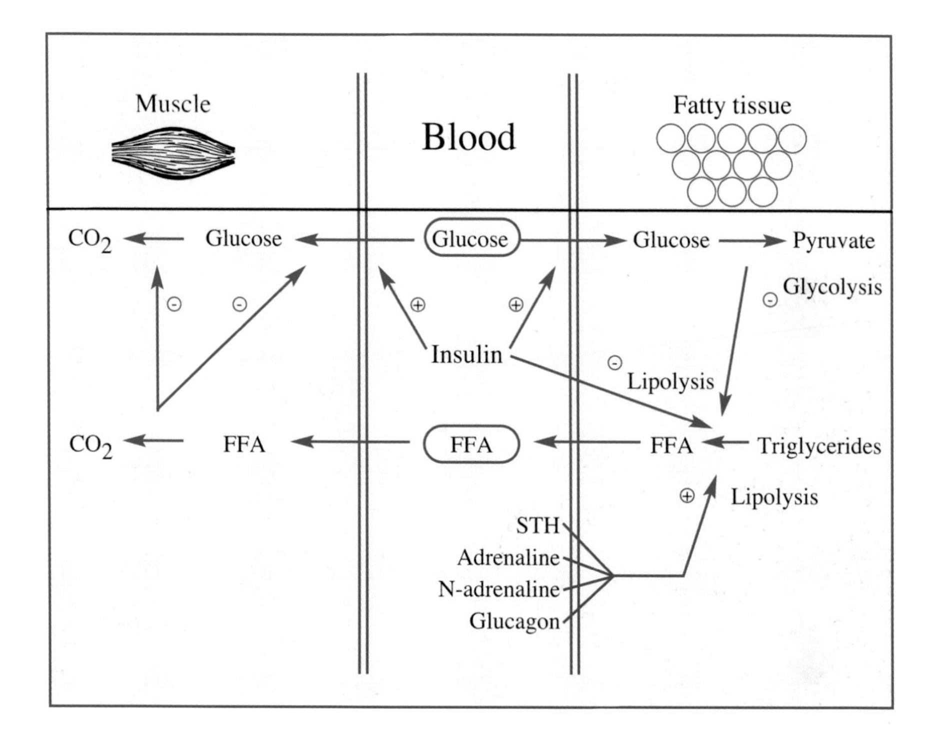

Fig. 4: Influence of hormones on glucose and fat metabolism. Insulin supports the flow of glucose into the muscles and also into the fatty tissue (in cases of excess). The growth hormone (STH), adrenaline, noradrenaline and glucagon support the release of free fatty acids (FFA) from the fat stores. High FFA concentration hinders glucose utilisation in the muscle.

Proteins have varying biological valency. **Biological valency** is an expression of how many per cent (%) of 100 g of the particular protein is utilised in metabolism (Table 2). Whole egg protein has a value of 100%. By combining egg protein with potato protein a biological valency of over 100% can be reached. Generally animal proteins have a higher biological valency than vegetable proteins. In competitive training **vegetarians** are at a disadvantage because through the plant proteins ingested the biological valency of animal proteins is not achieved. By combining various plant proteins the biological valency of food can be increased. In addition, vegetarian oriented athletes can take protein concentrate supplements.

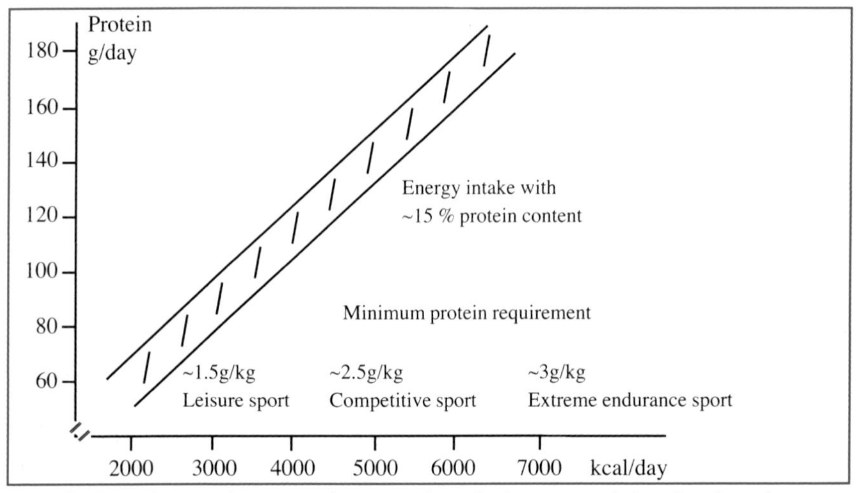

Fig. 5: Protein requirements in sport in relation to training load and energy ingestion

Biological Valency of Protein Mixtures in Nutrition

Protein	Biological Valency (%)
Whole egg	100
Meat	95
Fish	94
Milk	88
Cheese	85
Soy beans	84
Rice	70
Bread	70
Potatoes	70
Wheat	56
Corn	54
Beans and corn	101
Milk and wheat	105
Whole egg and milk	122
Whole egg and potatoes	137

Table 2

2.4. Water and Mineral Balance

50-60% of an adult's total body mass consists of water, with variations of 45-70% (Fig. 6). At 70 kg body mass, 60% water represents a water balance of 42 l. The water is distributed amongst three areas (compartments):

Blood plasma (3-3.5 l), intracellular space (8 l) and cell nucleus (30 l). In the muscles of a 70 kg athlete there are about 22 l of water.

As a result of physical load the water content is redistributed among the compartments. Major water loss via sweat leads to **dehydration**. Water loss in the compartments differs at rest and under load. Whereas **water loss in the sauna** is mainly from the intracellular space, during physical load or **active sport** it is from the intracellular space and the cell nucleus. Between the tissues water exchange is regulated by osmotic forces which are maintained by low molecule proteins and electrolytes (Table 3). In the body movements of water are always accompanied by mineral changes because the water and mineral balances work together functionally. Even when there is great dehydration the water balance in blood plasma is kept constant for a long time; only at about 5% water loss does the osmotic effect of blood proteins lead to a depositing of **water in the blood vessels (hypervolaemia)**. This **hypervolaemia ("blood**

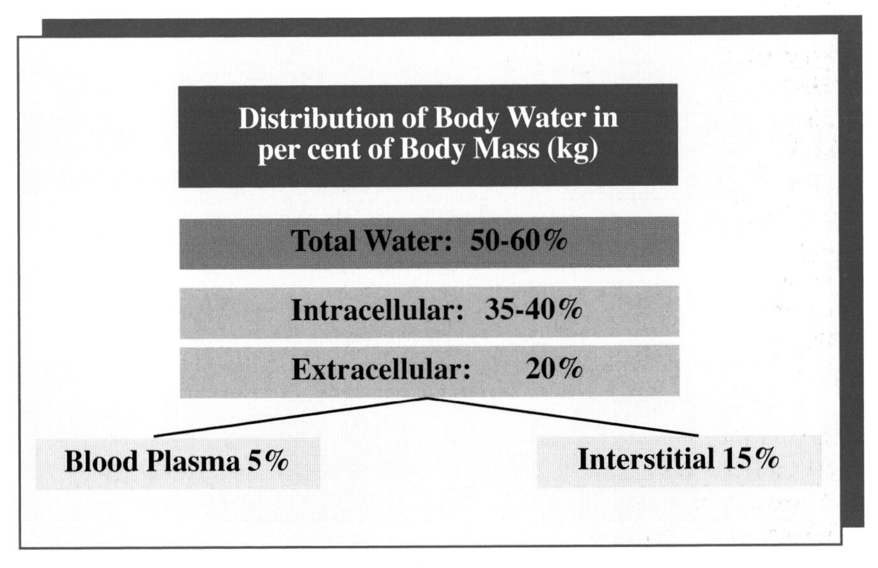

Fig. 6: Distribution of water in the body spaces

dilution") is a necessary regulation for securing the supply of oxygen and energy during long duration load. The opposite form of regulation, **blood thickening** or **hypovolaemia**, usually only occurs during short, intensive loads such as ergometric stage tests.

With the release of sweat **minerals** are excreted. Mineral loss increases with increasing volume of sweat. The salty taste of sweat indicates the excretion of salt (NaCl). Regular, heavy sweating is a cause of the loss of certain minerals and electrolytes (Table 4). Those minerals which are dissolved as salts in anions and cations are called electrolytes. The known **electrolytes** are sodium, potassium, magnesium and calcium.

In the case of sporting loads up to 30 minutes duration, measurable weight loss results mainly from **water loss**. After 10 to 30 minutes of intensive load in a warm environment, sweat loss can be 1-2 l. With longer load the weight loss is coupled with substance breakdown and water loss. In every hour of load, 200-250g glycogen and triglycerides (fatty acids) are broken down. The glycogen stores in the muscles and liver are sufficient for about 120 minutes of load. Because with one gram of glycogen about 3 g of water are released, the measurable loss of mass is greater. The water solution is partly used in metabolism. After glycogen breakdown is finished, mass loss slows down because 60 to 70% of energy then comes from the combustion of less water rich fatty acids with a higher energy content (see Chapter 5).

In cases of major **dehydration** body mass can be reduced by over 8% (Table 5). Major dehydration reduces performance and leads to pronounced functional disturbances.

Osmotic Pressure in Tissues

Degree of pressure is osmolality (mosmol/kg)

Resting value:	280 to 295 mosmol/kg
Load:	310 to 320 mosmol/kg

Dissolved particles in body fluids have an influence on osmotic pressure, mainly on proteins, electrolytes (sodium), urea and glucose.

Table 3

In a warm environment major dehydration leads to an earlier overheating of the body (**hyperthermia**). More on this can be read in Chapters 2.5 and 5. The degree of dehydration does not automatically influence performance capacity. What does have an influence is the speed of dehydration. Slow water loss can be dealt with better than fast loss. A **water loss** of 2-4 l/day is normal in training and is compensated again over night. Only after dehydration has reached more than 5 l/day is it not possible to compensate the water balance in 24 hours. The body's water balance is regulated hormonally.

Anorganic and Organic Components in Sweat

Anorganic Components	**Unit**	**1 Litre**	**3 Litres**	**20 Litres**
Sodium	g	1.2	3.6	24
Chloride	g	1.0	3.0	20
Potassium	g	0.3	0.9	6.0
Calcium	g	0.16	0.48	3.2
Magnesium	mg	36	108	720
Sulphate	mg	25	75	500
Phosphate	mg	15	45	300
Zinc	mg	1.2	3.6	24
Iron	mg	1.2	3.6	24
Copper	mg	0.06	0.18	1.2
Organic Components				
Lactate	g/l	1.5		
Urea	g/l	0.7		
Ammoniac	g/l	0.08		
Vitamin C	g/l	0.05		
Carbohydrates	g/l	0.05		
Pyruvate	g/l	0.04		

Table 4

Fluid Loss and Performance Capacity in Sport

Reduction from starting weight	Mass Loss from 70 kg	Performance Capacity and Symptoms
1 %	0.7 kg*)	Full performance; thirst
2 %	1.4 kg	Performance maintained with great effort
3 %	2.1 kg	Performance drop ~ 5 %, major fatigue
4 %	2.8 kg	Performance drop ~ 10 %, some stopping
5 %	3.5 kg	Performance drop ~ 15 %, exhaustion, high rate of stopping
6 %	4.2 kg	Performance drop ~ 20 %, muscle cramps, disturbances of movement coordination
10 %	7.0 kg	Break off in performance, reduction of blood flow through the kidneys and urine production by 50 %, disorientation, coordination disturbance, somnolence
15 %	10.5 kg	Unconsciousness, danger to life, death possible

**) About 85 % of mass reduction is fluid loss*

Table 5

The **hormones regulating** the water and electrolyte balance are aldosterone, atrial natriuretic peptide (ANP), antidiuretic hormone (ADH) and catecholamines. To protect the body against water loss the hormones increase during load. The effect of hormonal regulation of water and electrolyte movements in the body is that only little water is stored. Excess water taken in is immediately excreted again, something experienced by everyone when they drink a lot. Very diluted (light coloured) urine is a sign of excess water.

Thirst arises when fluid loss is over 2% or when very salty food is eaten. Compensating a fluid deficit with low mineral tap water is disadvantageous for athletes (see Chapter 5.1). Consuming water with a **low mineral content** extracts sodium from the blood in order to absorb the fluid.

It should be noted that many drinks containing sugar, such as lemonade, tea, cola etc. are only mixed with tap water and are therefore low in minerals. In a dehydrated state it is best to drink mineral water for regeneration, supplemented with fruit juices and fruit. Increasingly at major sporting events only tap water is provided along the course and at the finish. This is good for cooling the body, but not for general refilling of a fluid deficit. Alcohol is also not a drink for regeneration of a fluid deficit.

2.5 Temperature Regulation During Exercise

Independent of the outside temperature, sporting exercise lasting several hours leads to an increased body core temperature. The **body core temperature** rises to 39-40°C. An increase in body core temperature can regularly be established after marathon runs. Vigorous exercise in hot conditions causes the core temperature to rise even higher. At 40°C, however, a limit is reached above which serious effects on health are possible. This overheating of the body **(hyperthermia)** is highly dangerous (Fig. 7).

The degree of **heat development** is dependent on several factors. These include the size of the muscle used, the duration of load, the intensity of load, general performance capacity, as well as the sport specific training state. Duration of load has a great influence on the development of heat and the use made of the regulatory systems for the release of heat. The **mechanisms of heat release** are conduction, convection, radiation and evaporation (of sweat) (Table 6).

Heat conduction occurs according to the laws of physics: heat is conducted from a place of higher temperature to places of lower temperature. In this way e.g. cold hands get warm after a few minutes of exercise. Thick subcutaneous fatty tissue is a hindrance to heat release; overweight people sweat more easily because they are less able to release their body heat. Thick subcutaneous fatty tissue insulates well.

Heat convection in the body is supported by circulation to the tissues. The blood transports the heat from the muscle tissues to the skin surface and from here it is radiated. As air currents around the body determine the release of heat from the skin surface, sporting exercise in slightly moving air is more pleasant than when there is no wind.

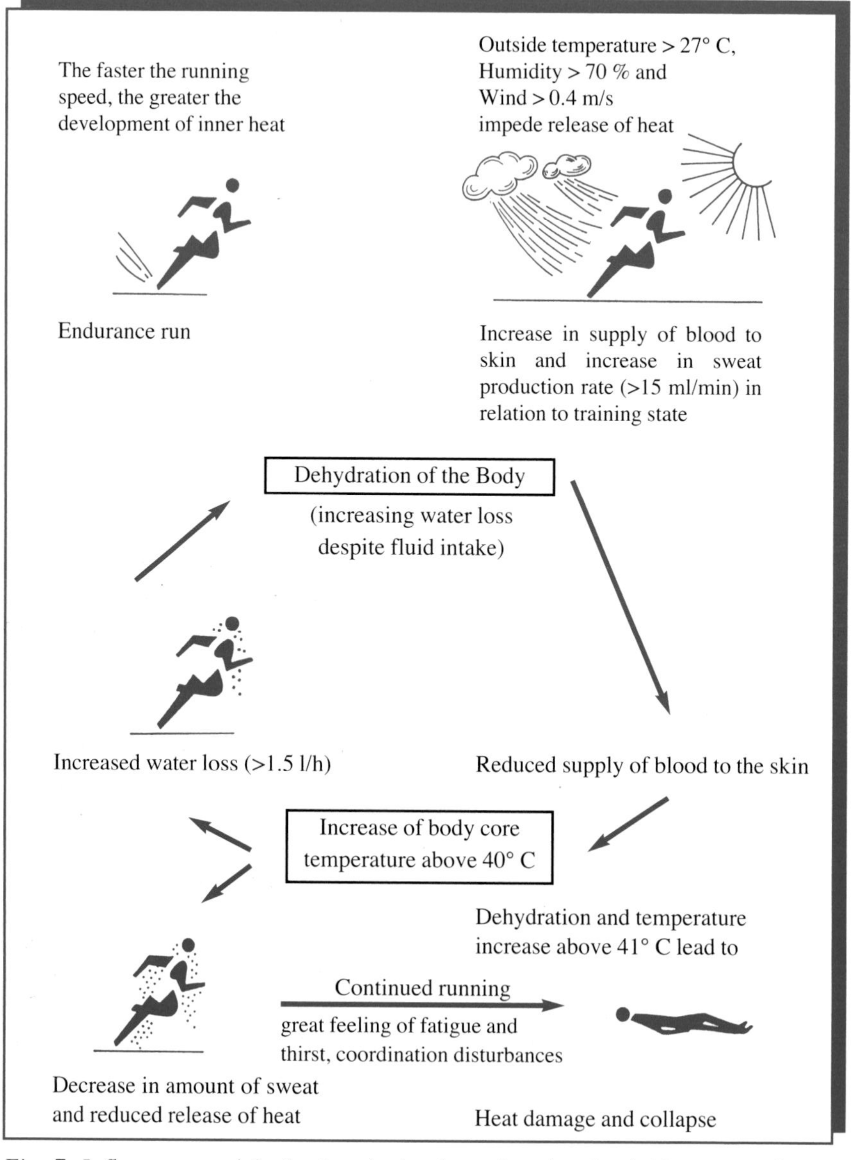

Fig. 7: Influences on dehydration during long duration load. No, or insufficient, drinking when running in heat leads to an increase in body core temperature, major dehydration and the danger of collapse as a result of heat damage.

Mechanisms of Heat Release During Sporting Exercice

Forms of Heat Release	Proportion of Heat Release [%]
Conduction	10 - 15
Convection	10- 15
Radiation	10 - 15
Evaporation	60 - 80

Table 6

Heat radiation involves the mechanisms of heat release via radiation through the skin on the infra-red wave length. This form of heat release only works at outside temperatures over 35°C. If the radiation temperature of the sun is higher, the skin is heated up. This effect is used in sunbathing at the sea, whereby after warming up, bathing in the cold water can be coped with better.

Sweat evaporation is the most effective way to remove heat dammed up in the body. The cooling effect does not come from the sweat itself, but from its evaporation. Evaporation draws heat from the body when the sweat changes from a liquid to a gas state. Hereby 1 g of sweat draws 0,6 kcal of heat from the body. When about 20 g of sweat per minute is developed during intensive load, this means heat extraction of 12 kcal/min or 720 kcal/h. Together with simultaneous evaporation of moisture in the breath this means heat extraction of about 800 kcal (3.352 kJ) per hour. In high humidity (over 80%) evaporation is more difficult. In high humidity sweat drips off the body and can thus no longer cool it. The greatest cooling effect comes from fine sweat spread over large parts of the body. Great loss of sweat not only disrupts the water balance but also the mineral balance.

The most effective preventive measure during exercise in heat is to drink fluids (with electrolytes) early and in large amounts. Fluids should be drunk before, during and after exercise.

3 Forms of Nutrition and Energy Intake in Sport

3.1 Nutritional Risk Groups

Regardless of the possibility of optimising energy ingestion according to demands, there are sports in which the athletes' views on norms regarding nutrition vary considerably (Table 7). Often the sport practiced, or person's occupation force people to adopt dietary behaviour which, if not adhered to, can reduce the chances of being able to carry out the occupation or achieving high sporting performance capacity. Occupations such as a modelling, ballet dancing, acting and figure skating or competitive sport gymnastics and acrobatics can only be carried out over longer periods of time through **restrictive dietary behaviour**. A different situation is found among **body builders**, who strive for muscle growth through special energy and protein ingestion. The third extreme group with regard to nutrition consists of the **long duration endurance sports**. These are the athletes who can only maintain hours of performance through constant ingestion of food during exercise. They have to harmonise food consumption, recovery breaks and sleep with exercise. The fourth extreme group are the **weight class sports**. Through temporary and repeated weight reduction athletes try to achieve success in a lower weight class.

3.1.1 Maintaining Low Body Mass

In particular girls and young women in the technical-compository sports such as gymnastics, rhythmic sports gymnastics and figure skating are faced with the problem of maintaining low body mass. Female dancers too, especially in ballet, must ensure they have low body mass for many years in order to stay in their occupations. Neither for athletes nor professionals are there any secret recipes for maintaining low weight. The key to success is **maintaining a low calorie diet** together with several hours of physical training daily. Permanent "starving"results in the fat reserves becoming very small. The proportion of fat is usually under 5% of body mass and thus well under the values of sport practitioners. Total metabolism adjusts to this state in an economising way, resulting in a kind of hunger metabolism without great energy reserves.

Risk Groups in Energy Consumption in Competitive Sport

Advantages	Sports	Type of Diet
Maintenance of low body mass and securing capability of exertion.	Gymnastics, ballet, rhythmic sports gymnastics, figure skating	Long periods of diet with too few calories are compensated for with protein concentrates, vitamins and minerals
Build-up of large muscle mass with little fat	Body building, weight lifting	Protein rich diet (up to 3g/kg). Deliberate ingestion of arginine, ornithine, glutamine and tryptophan to build up muscles.
Development and maintenance of muscle strength endurance, securing of muscle regeneration after great physical stress	Cycling, running, swimming, rowing, canoeing, triathlon etc.	Doubling of energy intake (4500-6500 kcal/day). Increased protein ingestion (12-17% of nutritional energy or 1.5 to 2.5 g/kg.) Supplementation of carbohydrates, vitamins, minerals and active substances for the development of performance and ensuring regeneration.
Rapid change in body weight, participation in lower weight class	Wrestling, judo and boxing	Repeated drastic change between energy and fluid ingestion. After dietary restricton, deliberate ingestion of fluids, carbohydrates and proteins in a brief space of time

Table 7

Dietary market

1.	Low salt food and beverages
2.	Low calorie food and beverages (light...)
3.	High calorie food
4.	High protein food
5.	Low fat food
6.	Dehydrated fruit juices and fruit
7.	Food with high fibre content
8.	High vitamin and mineral food
9.	Special additional food for athletes (high carbohydrate or protein, with or without minerals and/or vitamins)
10.	Active substances (L-carnitine, amino acids, vitamin Q etc.)

Table 8

For performance capacity, daily food ingestion in small servings is essential. This fills up the liver glycogen which maintains the blood sugar concentration. Going hungry for several days reduces concentration and performance capacity considerably. This state is very risky for exercises requiring high concentration. Carbohydrate ingestion, even if in small amounts, must be spread over the whole day in order to maintain the blood sugar concentration. This is a key prerequisite for securing the cerebrum's overview regulation and thus physical load tolerance. The plentiful variety offered by the dietary market (Table 8) can be used to support this.

Foods with high valency components (minerals, vitamins) and nutrient ratios useful for sport

Grains
Bread, pasta, rice, oatflakes

Milk
Dairy products (yoghurt, curd cheese, cheese)

Fruit
Bananas, citrus fruits, apples, dried fruits (raisins, figs, apricots)

Fish
Herring, trout, mackerel, salmon etc.

Beverages
Mineral water; apple, citrus, grape or currant juices; stout; beer (also non-alcoholic)

Nutrient mixtures
Muesli with grain germ, oatflakes, raisins, nuts, yeast etc.
(L-carnitine, amino acids, vitamin Q etc.)

Table 9

Limiting daily **total energy intake** to 1,200 to 1,700 kcal requires **wholesome nutrition** (Table 9). With this kind of diet, concentrated carbohydrates (monosaccharides) as well as white sugar, jam, chocolates, cake, biscuits and pastries, lemonade, cola beverages etc. should be avoided as much as possible (Table 10). These foodstuffs have a low nutrient density, i.e. they contain too few minerals and vitamins. Preference should be given to foods with a high **nutrient density**, such as wholemeal bread, fruit, vegetables and mineral water. In order to avoid the development of fat, a **spread of energy providers** of 40% carbohydrates, 33% protein and 27% fats is recommended.

Food should be high in protein and low in fat. Lean meat, eggs, low-fat cheese, low-fat curd cheese and low fat milk should be the preferred **main protein suppliers**. To make this low calorie diet psychologically bearable, small servings should be spread over six meals a day. Frequent food ingestion in small servings suppresses the feeling of hunger. Metabolism is regulated to absolute economy. The **vitamin and mineral** balance is essential to secure performance capacity and load tolerance. In this respect the ingestion of additional protein concentrates or amino acid mixtures with Vitamin B_6 (food supplements) is justified. In this way regeneration is supported in muscles as well as tendon and cartilage tissues that have had high demands placed upon them. In particular when girls are at their **growing age**, in the period of puberty, it is important to ensure **high protein ingestion**. This is important for ensuring growth and at the same time also of load tolerance at school and in sport. High training demands during puberty can lead to considerable delays in development, with body height and mass being well under the average development rate of the normal population (FRÖHNER, 1993).

Restrictive energy ingestion is often accompanied by reduced capacity to take load of the supporting and moving system of the body. With very slim girls disruptions to hormonal regulation are possible, resulting in irregular periods, weaker periods or no mentruation at all (amenorrhoe). This changed hormonal regulation can also be accompanied by disruption to bone development, fostered by a deficiency of minerals such as calcium, magnesium and fluoride.

A low calorie diet negatively influences the immune system on many levels. Immunological adaptation through sport, expressed in increased functional performance of the NK cells and the T auxiliary cells, an increase in the ability of cells to multiply, and in phagocytosis performance an increased number of cytocines (messenger substances), increased receptor development amongst others, is impaired. The no longer completely functional immune system reduces health stability, especially in psycho-physical stress situations and in dealing with illnesses.

In nutritional medicine the term **"immune nutrition"**was coined for this state of nutritional deficit. This means that through deliberate ingestion of active substances in conjunction with a low calorie diet, stabilisation of the immune system can be achieved (Table 11).

In order to strengthen the immune system, additional ingestion of unsaturated fatty acids (olive oil, fish oil), amino acids (arginine, ornithine, glutamine, tryptophan) and in particular antioxidants (vitamin E, vitamin Q, beta carotene, selenium) is recommended.

Carbohydrates in Foodstuffs

Carbohydrates	Type	Found in
Monosaccharides	**Glucose** Glucose, dextrose **Fructose** Fructose (fruit sugar), laevulose **Galactose**	Confectionary, beverage supplement Honey, fruit Milk
Disaccharides	**Saccharose** Beet and cane sugar **Maltose** Maltose (malt sugar) **Lactose** Lactose (milk sugar)	White sugar, jam, confectionary, stout Milk
Oligosaccharides	**Maltotriose** **Maltotetrose** **Maltopentose** Maltose mixture **Dextrine**	High-tech sports energy drinks White bread, crispbread, rusk
Polysaccharides	**Amylopectin** **Amylose** Vegetable starch **Glycogen** Animal starch **Pectin, lignin** Cellulose	Potatoes, grain, Bread Noodles, bananas Meat, liver Fruit, vegetables, roughage from grain outer layers (indigestible carbohydrate)

Table 10

Support of the Immune System to maintain its capacity to function after high demands

Measures	Objectives	Substances
Food	• Restriction of energetic deficits • Removal of states of protein breakdown, (catabolism)	• Carbohydrate ingestion (concentrates) directly after load • Pronounced protein ingestion (amino acid concentrates)
Mineral and element ingestion	• Support for the development of specific immune proteins • Ensuring immunological stability	• Magnesium, zinc, iron, copper, calcium etc.
Vitamin ingestion	• Support for development of antibodies • Breakdown of oxidative substances (free radicals)	• Vitamins C, E, B-complex, beta carotene
Infection prevention	• Consumption of immunoglobulins • Stimulation of immune system	• Supply of antibodies (immunoglobulin by doctor) • Ingestion of vegetable immune stimulants

Table 11

Nutrition for Children and Adolescents Practicing Sport

Load/effort:	School, plus 2 to 4 hours training per day
Energy requirements:	2500 - 3500 kcal/day at 40 - 55 kg body weight
Nutrition:	60% carbohydrates, 15% proteins, 25% fats; plenty of vitamins and minerals as well as natural products (Fruit, vegetables, whole grain products, milk and dairy products, meat)
Poor nutrition:	***Fast food diet:*** Hamburgers, French fries/chips, ketchup ***Beverages:*** "Coca-Cola", "Red Bull", "Flying Horses", "Guarana"etc. Ingestion of ***"empty calories"***: Confectionary, savoury biscuits, lemonades etc. ***"Night eating syndrome":*** High evening food ingestion with lack of appetite in the morning

Table 12

Special attention must be paid to the possibilities of incorrect nutrition for children (Table 12).

Currently the media propagate an **ideal of slimness** to the general population which leads young people and women in particular to try and copy it. Slimness is equated with happiness - a fatal irony! The consequences of this slimness or youthfulness craze are serious psychological problems and ongoing eating disorders. Dietary disorders can lead to a loss of the drive to eat called **anorexia**. Because there is often also a psychological component, the term anorexia nervosa is often used in connection with this condition. Anorexia nervosa involves massive weight loss and permanent fear of putting on weight. Another form of eating disorder is insatiable hunger. In the case of this eating disorder, known as **bulimia**, vomiting is often induced after eating. In extreme cases this results in a process of self-destruction among girls and women aged from 12 to 30 due to a refusal to eat, which requires long term treatment and appropriate guidance (self-help groups).

With **males** maintaining low body mass this is not such a grave problem. Usually it is a case of small statured people who in sport take on roles such as a cox in rowing, or who take up the occupation of jockey. Normal growth and good muscle development, as with gymnasts or ballet dancers, can aid performance if certain body proportions are maintained. These people must also keep to the principles of high protein, carbohydrate emphasis and low fat diet so that body mass remains low.

3.1.2 Body Building and Aerobic Dance

The training objective of **body building** and **aerobic dance** is varying development of muscles for posing. In aerobic dance there are also numerous dancing and acrobatic elements. In addition to the preferred and well known body building there are also more gentle forms, such as **body styling and body shaping**. **Aerobic dance** is preferred by **women**; competitions are becoming increasingly sophisticated and varied.

Muscle development only results from static and dynamic forms of resistance-oriented training **(strength training)**. With a normal **balanced diet** only minor muscle increases can be achieved. Strength training will only lead to an increase in the volume of the trained muscles if it is coupled with high protein ingestion. Protein ingestion, however, must surpass the normal DGE recommendation of 0,8 to 1.2 g/kg by a factor of 3 or 4 for **muscle growth (muscle fibre hypertrophy)** to take place.

High Protein Foods

Amount in each case is 100 g

Food	Proteins (g)	Carbo-hydrates (g)	Fats (g)	Energy Content (kcal/100g)
Low fat cheese	38.0	3.0	2.0	167
Peanuts	27.5	15.6	44.5	495
Fatty cheese	26.0	24.0	30.0	375
Beans (dried)	26.0	47.0	2.0	260
Lentils	26.0	53.0	2.0	300
Peas	23.0	52.0	2.0	290
Veal (lean)	22.0	—	3.0	111
Pork (lean)	21.0	—	7.0	140
Beef (lean)	21.0	—	4.0	115
Almonds	21.0	14.0	53.0	620
Herring	20.0	—	17.0	155
Chicken (fat)	19.0	—	9.0	171

Table 13

Body building requires protein ingestion of 2.5 to 3.5 g/kg body mass per training day. For aerobic dance 1.5 to 2.5 g/kg is sufficient. This amount needs not be consumed constantly, but only during periods of muscle development training. Like any other competitive athlete, body builders have performance peaks, such as championships, stage appearances and comparisons. Only at these times do they need to display maximum muscle development or a perfect acrobatic programme. When preparing for competitions, body builders go on a special high protein low fat diet (Table 13).

The sign of a well trained body builder, apart from having well developed muscles all over, is "thin"skin. By reducing fat ingestion, a dissolving of the subcutaneous fatty tissue is encouraged. In women body builders, restriction of fat ingestion and other influences can lead to a reduction in breast size (fat reduction). In order to achieve top placings, cosmetic surgery then becomes necessary. Obviously there is an increasing trend towards more gentle forms of body building as well as aerobic dance. The presentation of muscles is supported by deliberate water withdrawal. For this purpose athletes consume large quantities of potassium and little salt. In the cells sodium is exchanged for potassium and they shrink because the water-binding sodium is missing. Additionally, before a competition, participants refrain from drinking. As a result of "thin"skin, fluid abstinence and eating low salt, high potassium food, the muscles, hypertrophied by training, can be displayed to better advantage.

Regulating **diet** with natural protein suppliers **(meat)** is difficult because after a short time a dislike of meat develops. For this reason body builders tend more towards dietary foods with a high protein content which are manufactured in great variety. Combined products containing vitamins, minerals and other active substances are useful. Numerous **protein and amino acid preparations** are available. They are especially effective when taken after strength training because the excess amino acids consumed can quickly reach the place they are needed, the muscles. Major muscle growth is only achieved through protein ingestion of about 3 g/kg body weight, combined with long term and effective strength training. The amino acids are selected in such a way that a muscle anabolic effect takes place (see Chapter 8.1). Intermediate products of amino acid metabolism are also used, such as the breakdown stages of leucine. According to American research, beta hydroxy beta methyl butyrate (BMB) is said to have a definite muscle building and strengthening effect (ARMSEY/GREEN, 1997).

As mentioned already, fat ingestion must be sparing. The quality of the fatty acids is the key factor. Fat in nutrition should consist of at least one third essential and unsaturated fatty acids so there are no metabolism problems.

The taking of anabolic steroids, growth hormones, growth factor IGFI, insulin and other banned substances by successful body builders probably accelerates the dissolving of subcutaneous fatty tissue and also muscle hypertrophy. Mixing these building-up substances is highly dangerous and recently led to the death of a body builder from Austria.

Because of the health risk and the violation of ethical principles in sport through anabolic abuse, these should be absolutely avoided. According to BEUKER (1992) the side-effects of anabolic abuse are:

- Development of steroid acne in both sexes
- Increase in breast size in men
- Atrophy of the breasts in women
- Expothalmus (bulging eyes) in both sexes
- Development of baldness in men
- Changes in body hair (pubis, upper lip of women)
- Shrinking of penis and testicles; growth of the clitoris
- Swollen muscles and loss of subcutaneous fat
- Extremely broad shoulders in women

The **side-effects** named here should really be enough to stop any sensible body builder from using these anabolic active substances, especially as black market products could be impure. Meanwhile the external indications of anabolic use (e.g. steroid acne, enlarged breast in men, etc.) now lead to points being deducted in body building competitions.

Stable and well formed muscle growth should only be striven for through long term special maximum and strength endurance training, coupled with planned physiological dietary measures.

3.1.3 Long Duration Performance Capacity

For some time now the **marathon run** has no longer been representative of the limit to human endurance performance capacity. Today millions of runners can easily cover the distance of 42.195 km in 3-4 hours. Because many athletes no longer feel really challenged by the marathon run alone, longer and longer distances are being run, and extreme variants tried out e.g. ultra runs, multiple ultra triathlons, long distance swimming etc.

The trend to extension of distances is not only found in running. Combined **extreme forms of load** are well known such as the ultra triathlon (Hawaii triathlon). Even the distances of 3.8 km swimming, 180 km cycling and a marathon run in the ultra triathlon, which are covered in 8-14 hours, no longer satisfy some athletes. The **ultra triathlon** is now done even by women in multiples of three, four and five. In 1997 the world best time in the five times ultra triathlon was held by German woman Astrid Benöhr (74 hours and 4 minutes). Men meanwhile carry out ten and twenty fold times ultra versions.

These forms of load over several days call for new ways of storing energy for securing performance. **Food ingestion** mainly takes place **during exercise**, less in the short rest intervals. Reports on the experience of extreme athletes indicate that often several attempts were needed to reach the performance goal. One of the hindrances was often the problem of nutrition in connection with extreme exercise.

Nutrition analyses of the "Tour de France", and the cycling race across the USA (>4,700 km), indicate that the athletes used up 7,000-9,500 kcal/day (SARIS et al., 1989; LINDEMAN, 1991). On the **Tour de France,** 50% of energy is consumed on the bike, 60% of it carbohydrates. 61% of fluids was also consumed while cycling. Carbohydrate ingestion during the **ultra cycling race** (crossing the USA) was even more extreme. Here the cyclists consumed 78% of energy during the 24 hours of exercise in the form of **carbohydrates**. In training the athletes had found their level at a carbohydrate ratio of 65%. Reports indicate that the limits of glucose and fructose digestibility were not mastered. The consequences of too high a fructose ingestion were gastro-intestinal functional disorders. Information about digestive disorders is usually kept from the public unless they were the direct reason for giving up. Those athletes who are not sufficiently prepared for extreme exercise have insufficiently trained their fat metabolism. With daily repeated exercise, regeneration of the glycogen stores is delayed, as a result protein catabolism increases. Often the only answer is the infusion of a glucose and/or protein solution to maintain the performance capacity of the extreme athletes who want to reach the finish no matter what the cost.

During extreme exercise, 40 to 60 g of carbohydrates per hour must be consumed. This amount keeps the blood sugar concentration at a minimum level of about 4-5 mmol/l (72-90 mg/dl). During short and intensive endurance exercises (75% of VO_2 max), carbohydrate ingestion of 32-52 g/h leads to an increase in running distance and cycling time of 18 to 22% (Fig. 8). Total energy use consists of the ratios of carbohydrate, fat and protein metabolism.

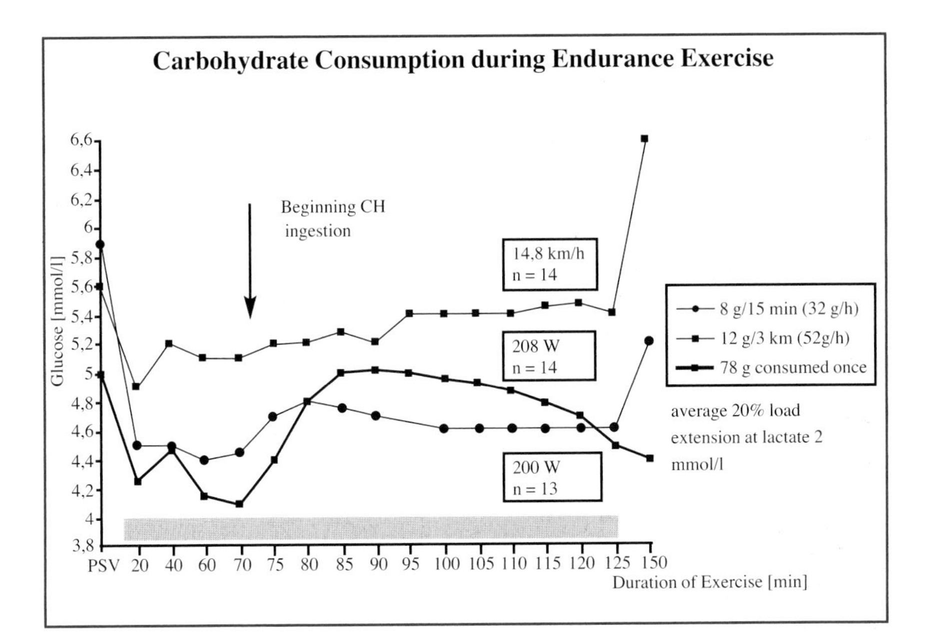

Fig. 8: Behaviour of blood sugar concentration during endurance exercise when various amounts of carbohydrates are consumed. Constant ingestion seems to be more advantageous than ingestion once only, and increases blood sugar less. Carbohydrate ingestion extends the duration of exercise by about 20% more than water ingestion (see Fig. 17)

In professional cycling a carbohydrate volume of 800-1,000 g/day, providing 3,280-4,100 kcal of energy, is needed even at low exercise intensity (30-40% of maximum oxygen intake). The other source of energy comes from the combustion of free fatty acids, which contribute 2,000 to 3,000 kcal/day. Theoretically another 1,025 kcal could be gained from the ingestion of 250 g of protein. From this it can be deduced that at daily exercise of 6-8 hours on the bike, total energy use of 5,900-6,200 kcal takes place. The peaks of energy consumption are the hill stretches on cycling tours, which require an average of 7,780 kcal/day and in some cases more (SARIS et al., 1989). In uninterrupted extreme competitions over several days (running, triathlon), energy consumption of about 500 kcal/h or 10,000-12,000 kcal in 24 hours can be expected. The duration of sleep alters energy consumption.

Training Load per Week

(Figures in km)

Type of Sport	Special Sports			Short Triathlon			
	Swimming	Cycling	Running	Swimming	Cycling	Running	Hours/ week
Leisure Sport	**5-10**	**100-500**	**30-50**	**2-5**	**30-60**	**20-40**	**4-6**
Popular Sport	**10-20**	**300-600**	**60-80**	**5-10**	**60-120**	**50-80**	**6-12**
Competitive Sport	**20-30**	**500-700**	**80-120**	**10-15**	**150-300**	**60-100**	**15-25**
Top Sport	**30-100**	**700-1000**	**120-250**	**15-20**	**300-600**	**100-200**	**30-40**

Table 14

With daily triathlon training of about eight hours a top athlete reaches total weekly load of 45-50 hours. In endurance sports, total energy consumption is dependent on the amount of weekly exercise (Table 14). If the amount of training is only 20 hours per week, which is the usual amount in competitive sport, then total energy consumption of 3,000-5,000 kcal/day must be covered by eating. The amount of carbohydrates needed for this consists of 50% simple and 50% complex carbohydrates and is 500 to 600 g/day altogether or 7-8 g/kg body mass. For energy consumption the actual time of exercise is what counts; taking longer breaks between sessions or sleep leads to false conclusions.

For this reason, at an energy consumption rate of 500-600 kcal/hour, **24 hour runners** who run continuously, or take breaks of about four hours, have **total energy utilisation** of 10,000-12,000 kcal. A disruptive factor during extreme exercise is the breakdown of the body's own proteins. During load protein synthesis is cut back. Both factors affect regeneration. The decrease in the development of new protein only makes itself noticeable several days after the exercise. The degree of muscle overload can be recognised in the increased creatine kinase activity (CK). Recent research shows that the increase in breakdown products of the slow twitch fibres (STF), the myosin heavy chain fragments, lasts a further 6-10 days. (KOLLER et al., 1994). Through protein breakdown for gluconeogenesis an energy gain from protein of 5-10% during exercise can be expected. This means protein breakdown of up to 5 g/hour.

If we calculate the pool of free amino acids with 100 g in the cells, and another 600 g of functionally moving specified proteins, the possible degree to which protein catabolism can disrupt body function during extreme exercise becomes clear. When there is major protein catabolism, **recovery** after long duration exercise can sometimes take **months**. The immune system can also be disrupted for a long time. Part of its stock of protein is probably metabolised to gain energy in energy-needed emergency situations. After extreme exercise athletes become ill more often than usual. Because of the exercise the immune system is depressed and the athlete is almost unprotected against pathogenes.

If a great need for extreme exercise is felt, in the interest of their own health athletes should maintain a careful nutritional regime after exercise as well. This diet must qualitatively secure the after-effects of exercise and support several weeks of recovery.

3.1.4 Frequent Changes in Body Mass

In combat sports, especially wrestling and boxing, frequent "weight adding"is a common procedure. The brief reduction in body weight takes place with the intention of fighting more successfully in the next lower weight class. The body weight is established through obligatory weighing the night or a few hours before the fight. Up until weighing the athlete makes use of various methods of dehydration and dieting. **Dehydration** can be achieved through vigorous exercise in warm clothing, by not drinking and by taking saunas. Dehydration using drugs (diuretics) is banned in these sports and is considered doping. Diuretic herbs are, however, permissible. All forms of rapid dehydration are risky because they can lead to a loss of muscular strength.

Dehydration of more than 5% of body weight disrupts important functional systems and leads to a loss of strength, especially of strength endurance performance above 30 seconds duration. Functional influences in the electrolyte and mineral balance are hard to overlook, usually the sodium and potassium balances are disturbed. The practical problem lies in the **rehydration** phase which occurs after weighing until the fight takes place. In 2-20 hours **refilling of energy and mineral stores** is possible. Infusions are also used here. This does not guarantee, however, that the original performance capacity of the muscles comes back and the trained maximum strength and speed strength capacity is completely available. Weight manipulation practice shows that unexpected **drops in performance** can always occur. In this situation weight reduction over a longer period of time seems to be the more favourable variant from a performance psychological point of view. To avoid disturbing performance capacity through reduction of body mass, **weight reduction** of 1 kg per week is recommended.

Of all the slimming diets, the so-called **"Formula Diet"** offers the most security because it contains all active substances (vitamins, minerals), even with reduced energy intake. If you opt for a different variant, mineral water, electrolyte carbohydrate beverages and carbohydrate protein concentrates should be given preference. Orientation towards high quality nutrients is important because some athletes carry out weight manipulation about 30 times per year. By doing so they apply major disruptive stimuli to the organs and functional systems. When taking off weight the energy content of food should be at least 500 to 1,000 kcal/day. Brief weight reductions are better tolerated if basic performance capacity is good.

Because of the great significance of **fast** weight reduction for health and performance capacity, FLEECK/REIMERS (1994) have recommended the following **rules:**

- When preparing for a competitive fight, highly nutritional foods should be consumed without reduction of fluid intake.
- Several days before the event all high volume foods (fibre) should be avoided.
- 48 to 24 hours before weighing, quantities of food and fluids should be reduced.
- After weighing, glucose electrolyte solutions should be given preference for rehydration.
- Excessive drinking is advised against.
- Complete rehydration takes over 24 hours.

3.2 Types of Nutrition in Sport Groups

3.2.1 Fitness Sport

Fitness or leisure sport is not a single sport group. Because of the large number of active participants and the diversity of exercise it is a central category in sport. Average training load is 4-8 hours per week. Because regular training is not compulsory, training restrictions or interruptions for occupational or other reasons are normal. The additional energy utilisation through sport is of personal significance for the prevention of cardiovascular and metabolic illnesses. Regular training with energy utilisation of 4,000 kcal (16,800 kJ) per week reduces the probability of a heart attack by 50% (PAFFENBARGER, 1982). Half of this amount has similar preventive effects.

The major effects of endurance oriented training include the breakdown of blood fats and thus the avoidance of the **risk factor of excessive weight**. **Leisure training** has the advantage that there need not be any major dietary restrictions. The degree of training load and the appetite after it regulate food consumption (Table 15). If body mass does not increase, then the balance is right.

Even with slow locomotion about 400-500 kcal are used per hour of exercise. If training is continued to an advanced **age** (above 60) and, if weight remains constant there need be no restrictions to eating habits. According to the recommendations of the DGE, with increasing age **energy ingestion** should be **considerably reduced** (Table 16). These recommendations are based on decreasing physical activity.

Energy Consumption in Leisure Sport 70 kg Body mass

Walking (fast march)	***10 min/km=6 km/h*** (25-30 km/week)
Duration of exercise (hours)	**Energy consumption** (kcal/h)
1	600
2	1200
4	2400
8	4800
Running	***5 min/km=12 km/h*** (30-50 km/week)
Duration of exercise (hours)	**Energy consumption** (kcal/h)
1	870
2	1740
4	3480
8	6960
Cycling	***2.5 min/km=20 km/h*** (100-200 km/week)
Duration of exercise (hours)	**Energy consumption** (kcal/h)
1	400
2	800
4	1600
8	3200
10	4000
Swimming	***30 min/km=2.0 km/h*** (4-8 km/week)
Duration of exercise (hours)	**Energy consumption** (kcal/h)
1	680
2	1360
4	2720
8	5440

Tab. 15

Recommended Daily Energy Ingestion According to Age

(German Association for Nutrition - DGE)

Age [Years]	15 - 18	19 - 35	36 - 50	51 - 65	>65
Energy ingestion [kcal/day]					
Men	3000	2600	2400	2200	1900
Women	2400	2200	2000	1800	1700

Table 16

The advantage of training at middle and advanced age is that almost the same amount of energy can be consumed as by young people without fear of excess weight. The processing of food consumed is not the same among people of the same age. Good "processors" must always load themselves more sportswise or physically in order not to put on weight. If a person with 80 kg body mass did 15 minutes jogging (2-3 km) every day, 240 kcal (1,008 kJ) would be utilised. If this energy consumption stops, a major increase in weight will develop over ten years. Experience shows that middle aged people who stop practising sport have excess weight of 5-15 kg. **Self-monitoring of body weight** allows early recognition of excess energy intake. **Reducing body mass** is possible with dietary measures (Fig. 9, 10):

- Orientation towards an energy reduced balanced diet of 1,200-1,500 kcal.
- Diet with industrially manufactured nutrient mixtures such as e.g. "formula diets"of 1,200 kcal.
- Potato or rice diet of 1,000 kcal.
- Outsider diets of 800 kcal etc.

One of the advantages of body mass reduction is that numerous risk factors are individually prevented or reduced. These include the preliminary stage of metabolic syndrome or disturbed glucose tolerance. The lowering of the glucose concentration in the blood leads to a decrease in carbohydrate pressure on the tissue. A constantly high blood sugar level fosters the development of fat. Food restriction lowers the blood sugar level (Fig. 10)

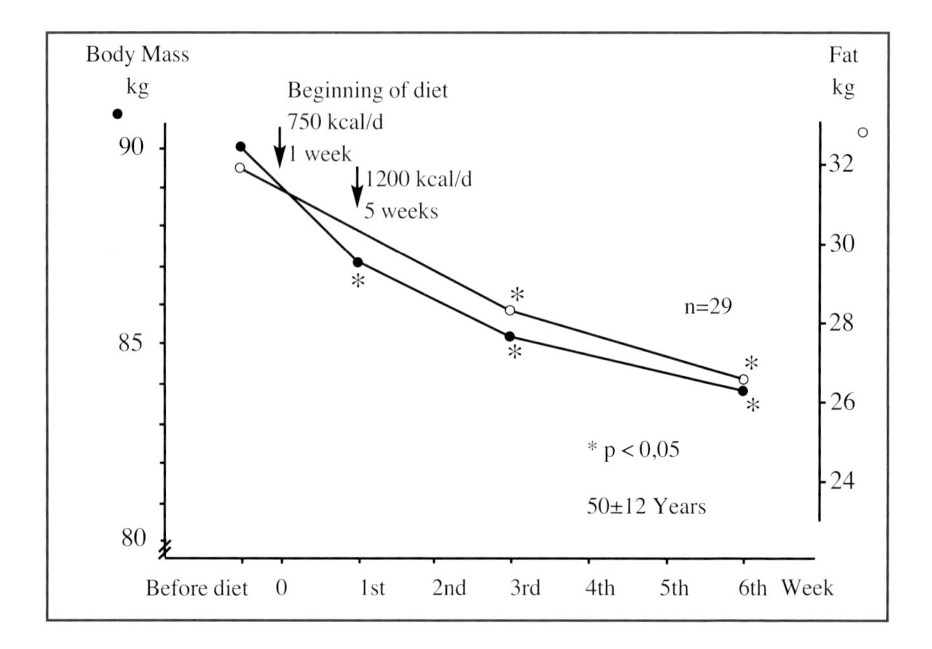

Fig. 9: Reduction of mass using a "Formula Diet", i.e. an industrially manufactured food product with vitamins, proteins and minerals, with 29 overweight people. After calorie reduction to 750 kcal per day in the first week, further energy restriction was more moderate (1,200 kcal per day). Body mass was reduced by a significant amount of 6.1 kg and fat significantly by 4.2%.

Even without deliberate weight reduction, leaving out sugar (e.g. in coffee) and/or restrictions to alcohol consumption result in reduced weight. Alcohol has a high heat or energy unit, 1g of alcohol burns 7.1 kcal!

In slimming procedures without sport, body weight can only be reduced if the amount of nutritional energy consumed remains below **basic utilisation**. Basic utilisation is, dependent on body weight, an average of 1,500-1,700 kcal (6,300-7,140 kJ). During seated activity there is additional performance utilisation of 500 kcal. Thus reduction of body weight can be achieved if energy intake is under the basic and **performance utilisation** of the day.

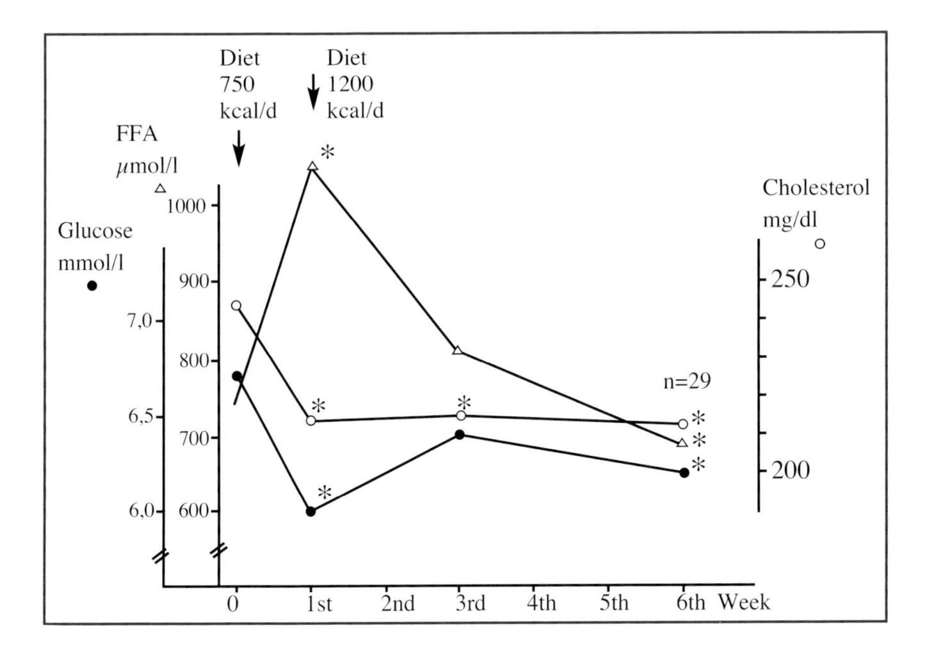

Fig. 10: Complementing Fig. 9, changes in metabolism measurement factors when food is reduced are shown. After a week blood sugar and cholesterol dropped significantly. Afterwards the concentrations remained at the same level. The Free Fatty Acids (FFA) rose considerably to start with because they were the replacement energy supplier.

With a daily **reduced energy intake** of 1,000 kcal (4,200 kJ), **weight reduction** of 1-2 kg per week can be expected, whereby the initial rapid drop in weight is always linked to loss of water. The real melting down of body mass is only about 0.5 kg. Producers of slimming formulas usually rely on the fluid deficit and declare this as real weight loss. The problem with successfully completed reduction of body mass is that after a year over 80% of people reach their former weight again if they have not drastically changed their lifestyle.

3.2.2 *Sport Groups*

a) Endurance Sports

Energy requirements in the endurance sports differ greatly because they are related to the duration and intensity of exercise, the application of energy (motion resistance) and motion technique. In endurance, energy provision is mainly secured through the body's **energy depots**. Because the energy stores are limited, additional energy intake is necessary during longer duration exercise. Daily **food ingestion**, with an average energy content of 2,000 to 4,000 kcal, provides the foundation for the securing of performance capacity in the sport. If energy requirements during longer training, or competition, load rise above the reserves of the glycogen stores, carbohydrates in the form of mono-, di-, oligo- and polysaccharides must be consumed during exercise (see Table 10).

Energy consumption increases during **endurance exercise** if training is carried out at increased sport-specific velocity and volumes (Fig. 11). The quality of the exercise depends on the training state. A criterion often used to estimate energy utilisation in sporting performance is the volume of **oxygen intake**. The volume of oxygen taken in one minute correponds to a certain **energy equivalent**. In carbohydrate combustion more energy is utilised than in fat combustion (Table 17). For rough estimates the energy utilisation from the oxygen used can be applied. When one litre of oxygen per minute is used it results in an energy equivalent of 5 kcal/min.

For example during exercise where in 60 minutes an average of 3 l of oxygen is taken in, 900 kcal of energy is used up (60 min x 5 kcal x 3 l = 900 kcal/h). This intensity of exercise can only be maintained for a limited time.

In average endurance training 2,000-3,000 kcal per day are used during exercise. Total energy use is of course higher, namely 4,500-5,500 kcal (Fig. 12). This represents training activity of 20 hours per week. If the volume is increased to 30-40 hours, energy use rises to 4,000-7,000 kcal. In modern competitive endurance training (cycling, triathlon), 6-8 hours of daily training is also normal. The necessary energy is ingested in the familiar ratios of the main energy providers, i.e. 55-60% for carbohydrates, 25-30% for fats and 10-15% for proteins. In studies involving competitive athletes these nutrient ratios are confirmed again and again (BROUNS, 1993; DIEBSCHLAG, 1993; EISINGER/LEITZMAN, 1992; HAMM, 1991; NOTHACKER, 1992; SARIS

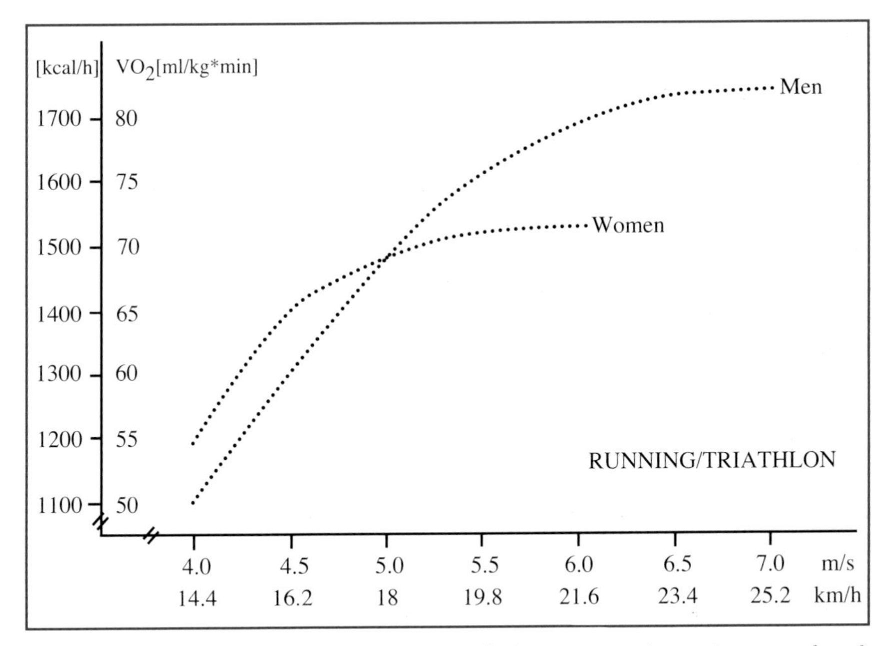

Fig. 11: Comparison of energy use in relation to actual running speed of competition oriented triathletes of both sexes. With increasing running speed energy consumption rose, but in relation to the level of aerobic performance (VO_2 max in ml/kg min).

et al., 1989 among others). **Female endurance athletes** also show no major variations (RÖCKER et al., 1992). They prefer carbohydrates with low nutrient density, however, such as cakes and confectionary. Practically the repeated call to raise the **carbohydrate proportion of one's diet** to 60% and above is difficult to comply with.

It means a diet that can even be practically vegetarian, or consuming carbohydrate concentrates in solid or fluid form as well as consuming mono- and disaccharides (KONOPKA, 1994). In competitions it is easier to eat high carbohydrate food than in one's usual diet. The greatest problem with this form of energy intake is ensuring a high **nutrient density**, the proportion of **monosaccharides** is between 45 and 60%. There is plenty of **saccharose** in confectionary, biscuits or cake. Cake and **confectionary consumption** almost reaches the level in bread of total carbohydrates consumed.

Energy Equivalent for 1 l O_2/min in Varying Metabolism Situations

Respiratory Quotient (RQ)		Type of Metabolism	Energy Equivalent (Oyxgen)
RQ	1.0	Only carbohydrate (CH)	5.05 kcal (21.2 kJ)
RQ	0.9	CH/Fats	4.93 kcal (20.7 kJ)
RQ	0.8	Fats/CH	4.81 kcal (20.2 kJ)
RQ	0.7	Only fats	4.69 kcal (19.7 kJ)

Energy Gain with Varying O_2 Intake and Assuming Energetic Equivalent of 1 l O_2/min as 4.9 kcal/min

1 l O_2 intake/min	equals	4.9 kcal/min (20.6 kJ)
2 l O_2 intake/min	equals	9.8 kcal/min (41.2 kJ)
3 l O_2 intake/min	equals	14.7 kcal/min (61.7 kJ)
4 l O_2 intake/min	equals	19.6 kcal/min (82.3 kJ)
5 l O_2 intake/min	equals	24.4 kcal/min (102.5 kJ)

Table 17

Numerous athletes consume up to 30% of their carbohydrates as monosaccharides with low nutrient density. If monosaccharide ingestion is too high, the risk of deficiencies of vitamins, fibre, minerals and trace elements (see Chapter 6) increases. This applies especially to the B vitamins as well as vitamin E and vitamin D. The other weak point of consuming too many monosaccharides is the deficit of iron, zinc, calcium and magnesium. If endurance athletes avoid complex carbohydrates and meat for longer periods of time, there is a high probability of suboptimal supply of iron, zinc and a number of B vitamins (Table 18). Targeted magnesium supplementation has now become common among many athletes so that deficiency situations are decreasing. A mineral deficit can occur again at any time, however, as has surprisingly also occurred among top athletes.

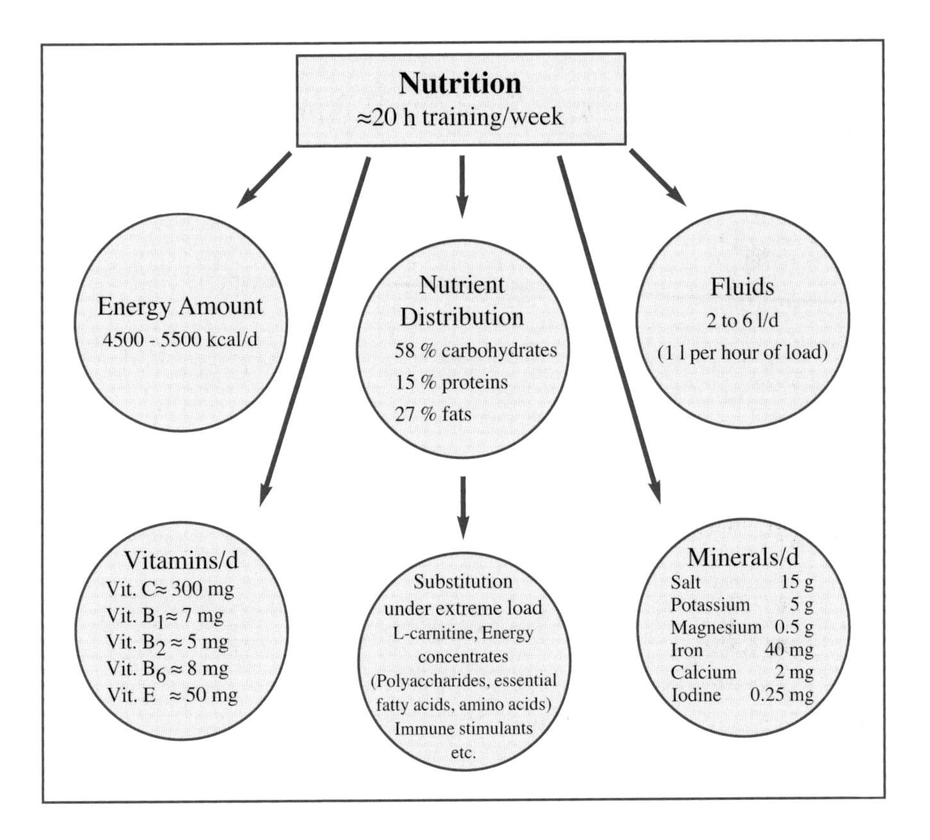

Fig. 12: Composition of a suitable sports diet for training load of 20 hours/week

b) Speed Strength Sports

Athletes in the speed strength sports represent extremes in nutrition because this group of sports includes the **maximum strength athletes** (weight lifters). In the speed strength sports e.g. long jump, high jump and sprinting, energy supply is not problematical. The main sources of energy for sprinters, jumpers and throwers among others are the breakdown of **creatine phosphate** and **glycolysis** (anaerobic glycogen breakdown). **Total energy consumption** in speed strenght sports is 68 to 72 kcal/kg of body mass (Table 19). The **ratio of energy supply** from carbohydrates is about 52%, from proteins 18% and from fats 30%. Thus energy consumption differs from the basic nutrition of other competitive athletes (Table 20).

Laboratory Readings of Iron Metabolism in Competitive Sport

Factor Measured	Unit	Norm Range
Iron	µg/dl (µmol/l) µg/dl (µmol/l)	60-160 (11-29) M 50-160 (9-29) W
Transferrin	g/l	2-4 M, W
Transferrin Saturation	%	16-45 M, W
Ferritin	µg/l 30-150 W	30-400 M
Haptoglobin	g/l	0,6-3,0 (after load <0.6)
Haemoglobin	g/dl (mmol/l) g/dl (mmol/l)	13-18 (8,1-11,2) M* 12-16 (7,4-9,9) W*
Haematocrit	%	40-52 M** 37-47 W

Table 18: M=men, W=women

* *After altitude training, increase of 1-1.5 g/dl! In cross-country skiing a racing ban of 14 days is imposed if values are >18.5 g/dl (M) and 16.5 g/dl (W).*

** *In cycling values above 50% result in a 14 day competition ban.*

The central factor is not total energy supply but rather protein ingestion. The physical appearance of sprinters gives an indication of intensive strength, or maximum strength training, which has a muscle building effect. In addition to a high ratio of protein, weight lifters must consume more fats because they need greater body mass to support the burdens they must accelerate (Table 21). Grilled poultry used to be symbolic of a weight lifter's diet, meanwhile this has been replaced by the ingestion of protein concentrates.

In weight lifting, rapid weight reduction often has to be carried out in order to gain an advantage by participating in lower weight classes. Because compulsory weighing takes place two hours before the event, manipulations of mass reduction must be completed by then. The measures for this are: going without food and drink, going to the sauna, consuming diuretic tea and laxatives. Consumption of **diuretics** is banned because they are on the **doping list**. Rapid loss of mass of over 2 kg through major dehydration can have a

Energy Requirements and Nutrient Ratios in Speed Strength and Maximum Strength Sports		
Energy Use	**Speed Strength Sports** (e.g. sprinting and jumping disciplines in athletics)	**MaximumStrength Sports** (e.g. weight lifting, throwing and shotput in athletics)
Total energy (kcal/kg body mass)	68 - 72	72 - 76
Carbohydrates (% proportion of energy)	50 - 52	45 - 48

Table 19

Basic Nutrition of Competitive Athletes (Nutrient Ratio)		
Carbohydrates (CH)	5-7 g/kg	≈ 55-60%*)
Proteins	1,2-1,8 g/kg	≈ 10-15%*)
Fats	1 g/kg	≈ 25-30%

*Table 20: *) Proportion of energy intake (calories)*
Over compensation: 7-8g CH/kg=65-70% of energy intake

negative effect on maximum strength ability. The safe way is **long term reduction of mass** through reduced energy intake which can be achieved in 2-3 weeks by reducing carbohydrates and fats as well as increasing protein in one's diet. During this period sufficient ingestion of minerals and vitamins must be ensured. Fluid volume should be at least 1.5-2 l/day so that the kidney function is not disrupted.

In weight lifting, normal competitive training always takes place at the upper limits of individual body mass, i.e. 3-5 kg above the competition weight aimed for. Because excess energy ingestion results in body fat, body mass must be checked daily.

High Fat Foods
Amounts are per 100 g in each case

Foodstuff	Fats (g)	Carbohydrates (g)	Proteins (g)	Energy Content (kcal/100 g)
Butter	84,5	0,5	0,8	785
Hazelnuts	63,0	7,0	17,0	670
Goose (fat)	44,0	-	14,0	445
Pork (fat)	34,0	-	16,0	362
Fatty Cheese	30,0	-	2,1	26,0
Eel	28,0	-	12,0	225

Table 21

In the putting and throwing disciplines (e.g. shotput, discus, hammer) these weight regulations do not apply. As a rule a big athlete with greater active body mass is at an advantage. The nutritional problem is in the ingestion of highly valuable proteins (amino acids) to secure specific strength performance capacity. In general these athletes consume over 2 g of protein/kg of body mass.

c) Combat Sports

The combat sports include the **weight class sports** judo, boxing and wrestling. In fencing there are no weight classes, here reaction speed is the key performance factor which is independent of constitution or body mass.

Chapter 3.1.4 contains notes on the problem of "making weight". In nutrition for one-on-one combat sport athletes the terms: **normal diet, reduction diet and build-up diet** have become accepted (Fig. 13).

The normal diet consists of 50% carbohydrates, 20% proteins and 30% fats. These energy percentages change considerably for the reduction or build-up of body mass, especially in the upper weight classes (Table 22).

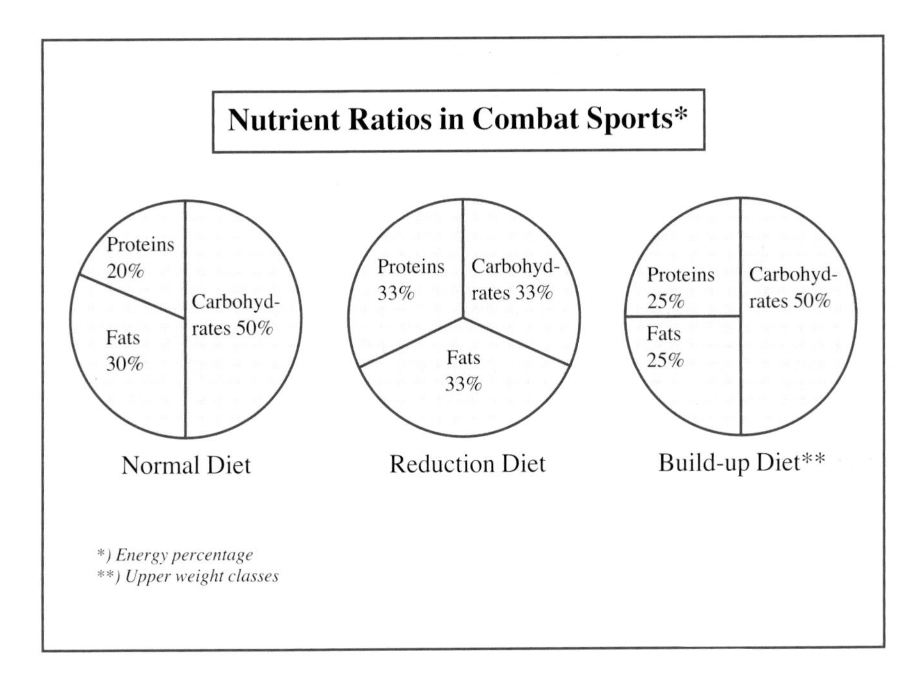

Fig. 13: Nutrient ratios in the combat sports for normal nutrition as well as reduction and increase of body mass.

Nutrient Amounts in Combat Sports

	Carbohydrates (g)	Proteins (g)	Fats (g)	kcal
Normal Diet	650	275	185	5458
Reduction Diet	115	115	55	1455
Build-up Diet*	1000	500	250	8475

*Table 22: * Upper weight class (85-110 kg)*

With regard to their performance determining abilities, the (one-on-one) combat sports have been classified as speed strength sports with high protein requirements (DONATH/SCHÜLER, 1979). In the middle weight classes (71-78 kg) the energy requirements are about 70 kcal/kg. This goes down by 10% in the lower weight classes (60-65 kg) and increases by 10-15% in the upper weight classes (86-105 kg). The orientation point is the competitive event. The normal diet is adhered to during training before an event in which a change of class is not intended. The **event diet** covers the time from weighing until the fight. The periods vary here: in judo, for example, the weight classes are determined 2-3 hours before the fight. In boxing weighing is also early, whereby the fight may not take place until the evening. In wrestling, on the other hand, weighing takes place on the evening before the day of the event. The influencing options of food and fluid ingestion vary accordingly. They are greatest in wrestling. The **structuring of diet between fights** is of major importance. Here it is advisable to ingest carbohydrates because "waiting stress" in the course of the day can lead to a major **glucose deficit**. This leads to unnoticed lack of concentration and technical mistakes during the fight.

In training and event breaks, maltodextrin drinks in concentrations of up to 15% have proved helpful. After events it is useful to consume high concentration glucose solutions and also amino acid mixtures, as these measures positively influence regeneration.

At adult level training in combat sports is done with body mass at 5% above event weight. For participation in the lower weight class, body mass must be reduced by 4-6 kg within a week. As already mentioned, it is recommended to gradually reduce body weight 3-5 weeks before the fight. **Supervised weight reduction** should be 0.2-0.3 kg/day and can be achieved through reduced energy intake, increasing the ratio of protein in food, and reducing salt. Drinking should be gradually reduced. The vitamin and mineral balance must be maintained absolutely. A zero diet is unsuitable in this situation. In order to make the feeling of hunger more bearable, eating should be spread over 5-6 meals. **Drastic fluid reduction** should only take place two days before the event. In training before this, 2-2.5 l should be drunk daily; this secures the functioning of the kidneys. Regular **self-monitoring of body mass** is advisable and also allows pre-calculation of the nutrients to be consumed. In addition, the **fat ratio** of body mass can be measured once a week. **Skin fold thickness measurements**, or other measuring systems for registering fat free mass are suitable for this. It should be noted that dehydration falsifies the measurements.

d) Game Sports

The commonly known **game sports** are soccer, handball, tennis, ice hockey, hockey, basketball, water-polo, rugby and table tennis. Game duration is e.g. 60 min (handball, ice hockey), 90 min (soccer), 60-120 min (volleyball) and 60-300 min (tennis). During this time, depending on the course of the game, players are loaded coincidentally (stochastically) and irregularly (acyclicly). Energy is gained in varying proportions from alactic, glycolytic and aerobic metabolism processes. **Speed** is energetically secured alactically and glycolytically, whereby creatine phosphate is significant energy-wise for specific manoeuvres in the game. The proportion of glycolysis varies. **Lactate concentration** is 3 mmol/l in volleyball, 4-6 mmol/l in soccer and over 10 mmol/l in ice hockey. For safe, and technically and tactically, stable manoeuvres a certain level of endurance ability is necessary, which is gained through game-specific basic endurance training. During continuous, intensive endurance training there is greater total energy use than in games which are interrupted by breaks. During the game energy consumption is 800-1.200 kcal/h. As athletes have a weekly load of about 15-30 hours of training, this results in **total energy requirements** of 4.500-5.500 kcal/day. In relation to body weight this represents 68-72 kcal/kg of mass. Tennis and basketball tend towards lower and soccer to higher energy consumption.

Game performance itself is mainly secured through the creatine phosphate stores and the glycogen stores via aerobic-anaerobic metabolism. Fat metabolism is partially significant for training. If there are high lactate concentrations as a result of too low game-specific aerobic performance capacity, inexactness in technical-tactical manoeuvres can occur.

Caffeine Content in Beverages

Beverage	**Quantity 100 ml**
Cocoa	≈ 10 mg
Cola	7 - 25 mg
Tea	50 - 100 mg
Coffee	60 - 120 mg

Table 23

As, however, concentration ability and coordination ability are mainly dependent on the **blood sugar concentration**; maintaining this is of major practical significance for all game sports. Drinking **fluids with glucose** in **rest intakes** is very advantageous for raising blood sugar concentration and partially refilling liver glycogen. 7 to 10 min after ingestion, glucose reaches the muscles and effectively works to improve performance during the game. Of the beverages usually available, normal cola is suitable because in addition to a sugar percentage of 10-12% it also contains small quantities of caffeine (Table 23). In most athletes, consuming caffeine through a cup of coffee stimulates the central nervous functions. During games additional vitamin or mineral ingestion is not necessary.

During tournaments, accelerated regeneration is important so that courses of action common to endurance sports are recommended (see chapters 3.2.2 and 3.3). This means that in the first hour of recovery concentrated carbohydrates should be consumed. Glucose drinks should be consumed even without feelings of hunger. The usefulness of taking protein concentrates during longer breaks in the game cannot be decided on by the experts.

Average Time Foods Stay in the Stomach

Hours	Foods and Beverages
1	Water, coffee, tea, beer, cola beverages, glucose*, carbohydrate solutions, amino acids, protein hydrolysates
2	Milk, cocoa, yoghurt, meat broth, rice, trout, carp, cakes, cream, bread rolls, white bread, tender vegetables, bananas, confectionery
3	Mixed grain bread, biscuits, buttered rolls, potatoes, apples, eggs, beef and mutton, chicken, vegetables
4	Sausage ham, turkey, roast veal, hamburger, fatty pork, nuts
5	Roasted poultry, game, legumes (beans, peas), cucumber salad, French fries
6	Bacon, herring salad, mushrooms, tuna
7	Sardines, roast goose, grilled knuckle of pork

*Table 24: *Glucose is in needy muscles within 7-10 min.*

In order to avoid straining oneself with digestive processes during the game, regardless of the time of day the last meal should be eaten 2 1/2 to 3 hours before the game. Attention should be paid to the digestibility of foods as they remain in the stomach for varying times (Table 24).

e) Technical Sports

Technical sports in the classical sense are gymnastics, rhythmical sports gymnastics, diving, figure skating and ski jumping. Without their acrobatic components alpine racing, sailing and shooting are also included here.

For most male participants in these sports nutrition is not a problem and does not differ much from untrained persons. The nutrition of **girls** in the technical-acrobatic sports, however, is a problem, especially in gymnastics and in rhythmical sports gymnastics. **Body weight** has a decisive influence on performance capacity, so that the world's best female gymnasts weigh on average only 43 kg and the men 62 kg. When the individual weight limit is exceeded, disadvantages in coordination performance and movements around one's own body axis occur. The practical difficulties for coaches lie in dietary guidance of girls in the developmental ages of pre-puberty and puberty. Here the energy requirements for growth and for load in biomotor learning training must be secured through appropriate dosages. All **growth surges** considerably disturb the training programme. **Nutrition** must take into consideration the requirements during **loading and unloading**. One day without training can cause problems because the risk of putting on weight occurs. Regulating body mass is not so complicated for boys as the performance age resulting from strength requirements in the training programme is later than for girls and women.

Total energy ingestion needed to maintain 35-45 kg of body mass is low, between about 1,200-1,900 kcal. The **nutrient ratio**, assessed in energy percentages, should be 40-50% carbohydrates, 20-32% proteins and 25-27% fats. Further details on maintaining low body weight can be found in chapter 3.1.1.

3.3 Carbohydrate and Protein Ingestion in Sport

Muscular performance capacity is tied to regular energy intake. The body's own **glycogen stores** enable performances of 90 to 120 min duration. Longer performances are possible through increased involvement of the **free fatty acids** in energy metabolism and additional re-developing of glucose **(gluconeogenesis)** from amino acids, glycerol and lactate.

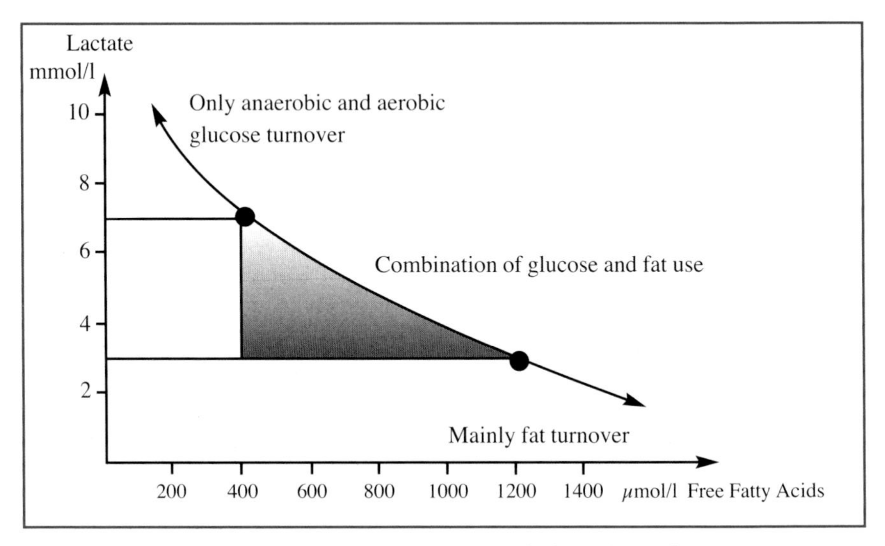

Fig. 14: Influencing of the ratios of energy metabolism through exercise intensity (level of glycolysis with lactate decrease).
Fat metabolism is only trained when lactate concentration is well below 3 mmol/l and duration is long lasting.

With increasing duration of exercise, even under intensive loads lactate concentration goes down and mobilisation of free fatty acids increases (Fig. 14). The drop in lactate concentration to below 3 mmol/l is the prerequisite for the unhindered processing of the fatty acids in metabolism. In order to maintain the blood sugar level, independently of this lactate-fatty acid ratio, carbohydrates must be consumed during exercise or in rest intervals. When food is consumed during exercise, not only the quantity is decisive but also its suitability for exercise metabolism. Today it is known that **carbohydrates are the most important nutrient**, which must be consumed **during longer lasting exercise**. This realisation is fairly recent. When e.g. previously in cycling energy was consumed during exercise of several hours duration, everyone concentrated first on foods with the greatest energy content, which included fats. Fatty broths among other things were fed to athletes during training or competitions. This energy intake showed little effect and was replaced by nutritional physiologically better suited carbohydrates (see Table 10). This was supported by experience in so far as it was discovered that so-called **"craving hunger"**is a form of hypoglycaemia and can easily be remedied by consuming glucose.

3.3.1 Carbohydrate Ingestion Before Exercise

Stored glycogen covers the energy required for sporting performance from a few minutes to as much as two hours duration. In sporting practice there have been varying findings regarding **carbohydrate ingestion** shortly before exercise. This applies to digestibility and effectiveness.

With greater endurance performance capacity, under similar loads **less glycogen** is required. Lower combustion of carbohydrates is compensated for by greater use of **free fatty acids**. In certain situations, especially when training is begun again after **illness** or **injuries**, the proportion of **fat combustion** goes down again. Energy compensation occurs through increased carbohydrate combustion.

Before important competitions, full glycogen stores increase performance. Better performances can be expected with normally or greatly filled glycogen stores than if they are not completely filled. Simply eating plentifully, without appropriate training, will not fill the glycogen stores above normal.

Glycogen over-compensation as propagated by HULTMAN (1974), with a special low carbohydrate and high fat and protein diet before competitions, is hardly practised any more in competitive sport. The reasons are as follows: to achieve glycogen over-compensation it is necessary to train for several days while consuming high protein and low carbohydrate food. During this time athletes feel weak and diarrhoea can also occur. At today's performance levels athletes can rarely afford an arbitrary training break of 3-7 days before important competitions, and from a sporting methodological point of view this measure is considered risky.

The alternative to classic over-compensation is to gradually reduce training load before important competitions and at the same time increase carbohydrate ingestion. A week before the competition daily load reduction takes place and parallel carbohydrates (CH) ingestion is increased. The quantity of CH should be increased from 6 to 10 g/kg of body weight per day. A total of 600 g/day, however, is enough; larger quantitities are of no additional benefit for filling the stores (BURKE et al., 1993).

Directly before a competition of over two hours duration a CH rich meal to fill up liver glycogen is helpful. Short competitions of under 60 min duration do not require CH ingestion. Intake of easily digestible CH in the form of energy drinks or energy bars, especially when there is little time before the start, is sufficient. It is also possible to participate in competitions of up to 90 min duration, the energy for which is covered by available glycogen reserves, without eating beforehand. The prerequisite for this is that the athlete feels good with an empty stomach and is mentally prepared to perform.

The advantage of training on an **empty stomach** is the **greater** proportion of **fat combustion** during load. Training on an empty stomach for a longer period of time is a way of "training fat metabolism". Muscle performance is not tied to the "fullness"level of the stomach as the muscles have sufficient **energy stores**. The direct energy stores for muscle work are creatine phosphate, glycogen and the triglycerides (neutral fats). With these the muscles have energy reserves for short or long lasting loads.

If **nothing is eaten** during a period of 12-6 hours before a competition, a **filling** of the **glycogen stores is not possible**. This can, however, still be corrected if short chain carbohydrates (glucose, maltose) in quantities of 200-350 g are deliberately consumed six hours before the start. For this reason **"noodle parties"** or more recently also **"potato parties"** before endurance events have become popular. These result in deliberate ingestion of carbohydrates ("carboloading"), but usually they take place the evening before the competition.

The **time of day** sporting demands take place is not unimportant for energy ingestion. If exercise begins early in the day, a different form of nutrition is required than for a later start.

Every form of food ingestion before a sporting event is advantageous as long as the level of "fullness" of the stomach does not hinder performance of the sport (see Table 24). The most common form of **pre-race food** in everyday practice is ingestion of industrially manufactured carbohydrate concentrates or bananas. These should be consumed two hours before longer lasting loads. Even directly before load carbohydrates (e.g. energy bars with 50-80 g of carbohydrates) can be eaten. Their ingestion has the advantage that after their absorption in the intestines the **blood sugar level** in the first part of load rises slightly and hypoglaemic regulation due to exercise is suppressed.

In numerous publications there have been warnings against **carbohydrate intake before load**. The objections of doctors and nutritional scientists to the practice of carbohydrate ingestion directly before loads are based on experience in hospitals with untrained persons. They were also partially based on research by COSTILL et al. (1977), who in laboratory experiments found a reduction in performance when glucose was consumed before exercise. They warned against hypoglycaemia due to carbohydrate ingestion. These findings have since been

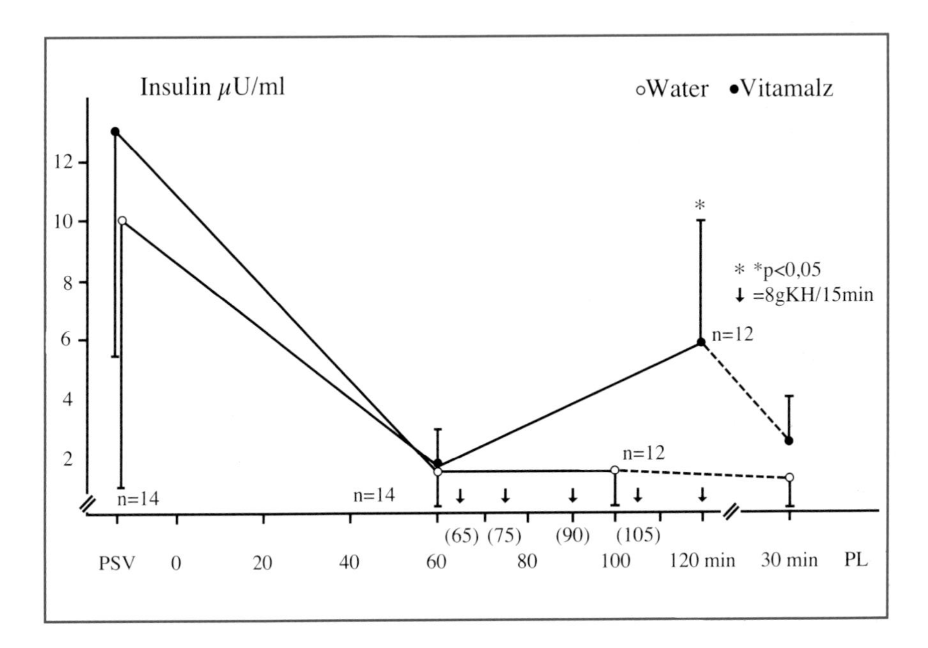

Fig. 15: Long duration cycle ergometer loading in the area of the aerobic threshold (lactate 2 mmol/l) with water and carbohydrate ingestion. The carbohydrate ingestion (Vitamalz(r)) of 32 g/h (arrows) only negligibly influenced the insulin which dropped during continuous load. After carbohydrate intake, blood sugar rose significantly ($p<0.05$) (see Fig. 17). PSV = Pre-start values. PL = Post-load.

refuted. In sport the feared **hypoglycaemia due** to glucose ingestion before exercise has proved to be non-existent. Checks on the results of COSTILL et al. (1977) by LEVINE et al. (1983) among others showed that in the case of **glucose ingestion before exercise** there is **no** drastic **increase** in **insulin** followed by hypoglycaemia. The increased release of insulin into the blood, as a result of glucose ingestion before or during exercise, is limited physiologically and is only minor in active athletes (Fig. 15). After 30 min of **endurance load** there is a major **drop** in **insulin concentration** below the pre-start value. In this metabolic situation glucose intake does not lead to a regulatory drop of the blood sugar level towards hypoglycaemia. The opposite is the case; in trained athletes, when **glucose is consumed** the **glucose concentration** in the blood always rises. Insulin concentration remains almost unchanged and considerably below the pre-start value. The increase in glucose is relative to the quantity consumed.

Too **great** an **ingestion of glucose or carbohydrates** before a start does, however, have a **disadvantage**. In the case of plentiful availability of glucose, **energy metabolism** adjusts to mainly **glucose combustion**. As a result, **fat metabolism** is temporarily **suppressed**. The working muscles burn the plentifully supplied glucose first. Glucose can be used in metabolism more quickly than fatty acids. The amount of glucose consumed before a race should, however, be limited, about 1 g/kg of body mass or 60 to 80 g. In the case of long duration loads, too great an ingestion of glucose before the start can lead **to early breakdown of glycogen (glycogen depletion)** because the use of fatty acids is suppressed.

The type of sport, or the intended load intensity, influence the volume of the carbohydrate mixture consumed before exercise. More can be eaten e.g. before cycling than before running. A full stomach is a great hindrance when running fast.

3.3.2 Carbohydrate Ingestion During Exercise

In numerous experiments it has been proved that **carbohydrate ingestion** during exercise maintains performance capacity and can extend **exercise duration**. In particular, the additionally consumed glucose has the effect of improving performance in the last third of an endurance exercise (COGGAN/SWANSON, 1992). Glucose ingestion does not lead to any further breakdown of muscle glycogen during exercise, as the authors mentioned proved through biopsy of the thigh muscles. The carbohydrates consumed increase the **glucose oxidation rate** and delay the timing of the drop in this. Thus the time of tiring is delayed at unchanged performance or velocity. **At least 30 g of carbohydrates** per hour of exercise should be consumed so that it has a measurable effect on glucose metabolism (BROUNS, 1993).

The consumed glucose is absorbed very quickly, and after about 7 minutes already starts to have an effect in energy metabolism of the muscle. Depending on their composition carbohydrates have varying influences on blood sugar level and energy metabolism. It has been proved that glucose, fructose and maltose (see Table 10) have similar effects on the raising of the level. Fructose, however, takes longer to be integrated in metabolism. It must first be broken down into glucose in the liver and then transported to the needy muscles via the bloodstream. Unlike other sugars, fructose ingestion has no influence on the release of insulin from the pancreas. For this reason for a long time it has been favoured in energy drinks. Not all carbohydrate solutions are equally digestible or tasty - factors which become significant to ingestion during exercise.

Solid and fluid energy ingestion in a training break

Cycling training in the hills of Teneriffe

Altitude training makes you thirsty.

A well earned break at the foot of the Teide at 2200 m.

A wetsuit protects against the cold, in swimming baths as well.

In comparison to glucose and fructose, **maltose** is very digestible. The small molecule size of maltose allows rapid absorption in the intestines and also ingestion of highly concentrated solutions (up to 15%). This finding is used in practice through ingestion of non-alcoholic stout during load.

The **multiple bond sugars (oligo- and polysaccharides)** also have advantages for ingestion during load. They are absorbed somewhat more slowly and take effect over a longer period because they must first be broken down into glucose. Muscles can only use **glucose directly to gain energy**. Preferred athletes' beverages contain mixtures of glucose, maltose and polyaccharides; in this way immediate and longer lasting effects are achieved.

There are varying views on the carbohydrates to consume during load, and especially on the timing of their ingestion.

Continuous carbohydrate ingestion during longer duration loads has proved to be a **physiologically effective form** for maintaining performance capacity. Consumption of carbohydrates (glucose, malt sugar among others) is recommended in quantities of 8 g to 12 g every 15 min, i.e. 32 g to 48 g per hour of load (COYLE et al. 1983; IVY et al. 1983; HARGREAVES et al. 1984 and 1987; NEUMANN/PÖHLANDT, 1994 among others). Consumption of a greater quantity of carbohydrates suppresses proportions of fatty acids in the transformation of energy. Therefore the **first ingestion of carbohydrates** should only **take place after about 60 min of exercise**. By this time the ratios of carbohydrates and of fatty acids in energy transformation have adjusted themselves to the level of performance reached, and at this time the glycogen stores of the liver are not yet exhausted. The gradual rise in the concentration of free fatty acids signals the increased share of the fatty acids in energy transformation (Fig. 16).

If **ingestion of carbohydrates** is **dosed** in quantities of 35 g/h, the regulated fat metabolism is not influenced (NEUMANN/PÖHLANDT, 1994). Consumption of 35 to 45 g of carbohydrates per hour during endurance exercise increases blood glucose concentration by 0.5-1 mmol/l (9-18 mg/dl). This **carbohydrate ingestion influences exercise duration**, which extends by about 20%; tiring of the muscles begins later (Fig. 17). Exercise is ended with a sufficiently high blood sugar level so that the causes of tiring are to be found elsewhere e.g. weakening of strength endurance.

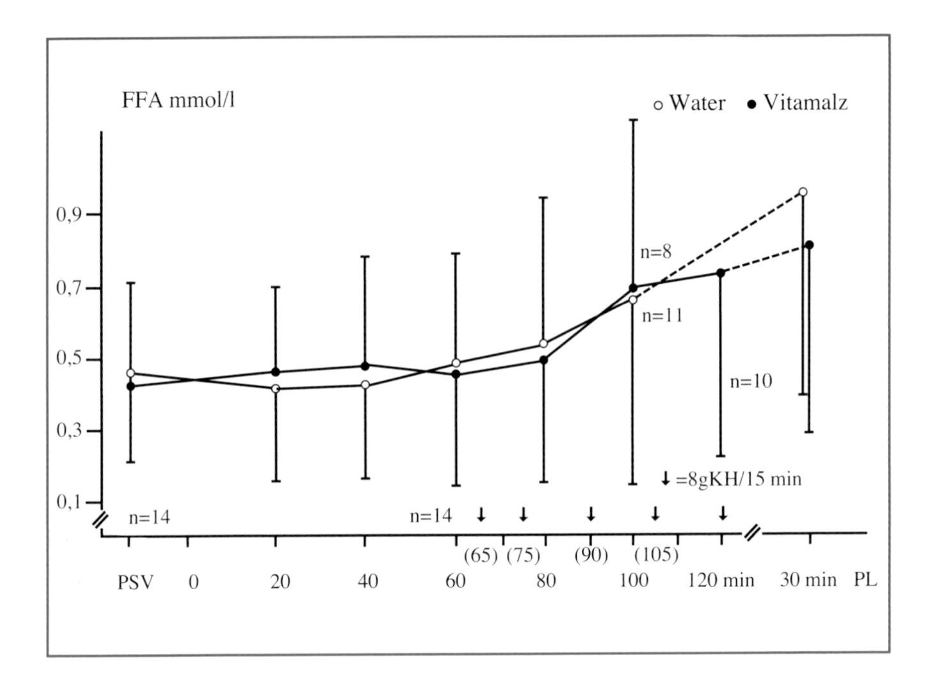

Fig. 16: Dosed carbohydrate ingestion during long duration cycle ergometry (closed circles) does not influence fat metabolism (free fatty acids = FFA) as opposed to mineral water ingestion (open circles).

In trained athletes who consume glucose or carbohydrates during load, **insulin** does not lead to hypoglycaemia. During long duration exercise the hormones **insulin and glucagon** always behave contrary to each another. It has been proved that after the 30th to 60th minute of an endurance exercise insulin concentration in the blood always sinks considerably. Glucagon on the other hand rises. Because of the glucagon increase, the free fatty acids are processed better, and gluconeogenesis extends stability of the blood glucose concentration. The Respiratory Quotient (RQ), the relationship between oxygen intake and carbon dioxide excretion, remains unchanged by dosed carbohydrate ingestion (Fig. 18). The conclusion is that ingestion of low percentage carbohydrate solutions does not influence fat metabolism.

During exercise the level of **blood glucose concentration** is determined by the amount of carbohydrates consumed. At ingestion of 30-40 g of

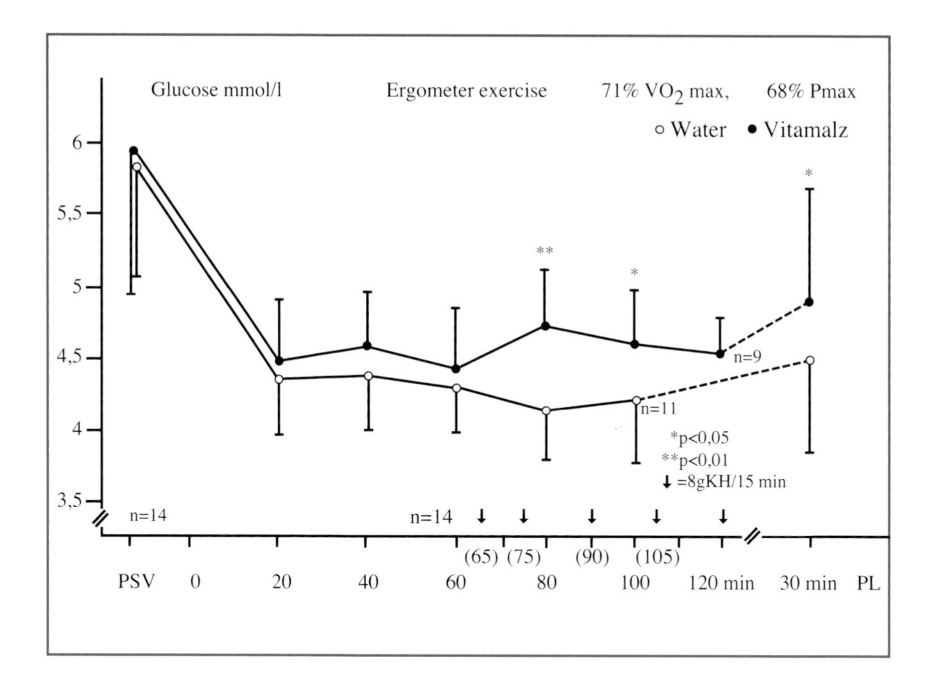

Fig. 17: Consumption of carbohydrates (Vitamalz ®) during cycle ergometer exercise in the area of the aerobic threshold leads to a significant rise in blood glucose concentration and extending of cycling time by 20%.

glucose, and long chain carbohydrates (polysaccharides), per hour of load the blood sugar level is always maintained at the necessary level for intensive endurance exercise. The state of muscle fatigue occurs independently of the blood sugar level, it is mainly determined by the level of training and aerobic performance capacity. **Hypoglycaemia**, however, caused by neglecting to consume carbohydrates during exercise of several hours always leads to reduced velocity, and finally to an **end of performance**. The ending of performance, or a great decrease in velocity of forward propulsion due to a glucose deficit, are caused by a failure of the motor driving forces in the cerebrum and cerebellum. The **brain always needs glucose** to maintain its performance, so do the motor driving forces. Consumption of carbohydrate mixtures during longer endurance performances does not influence, or only slightly influences, glycolytic metabolism, recognisable from the behaviour of lactate concentration (Fig. 19).

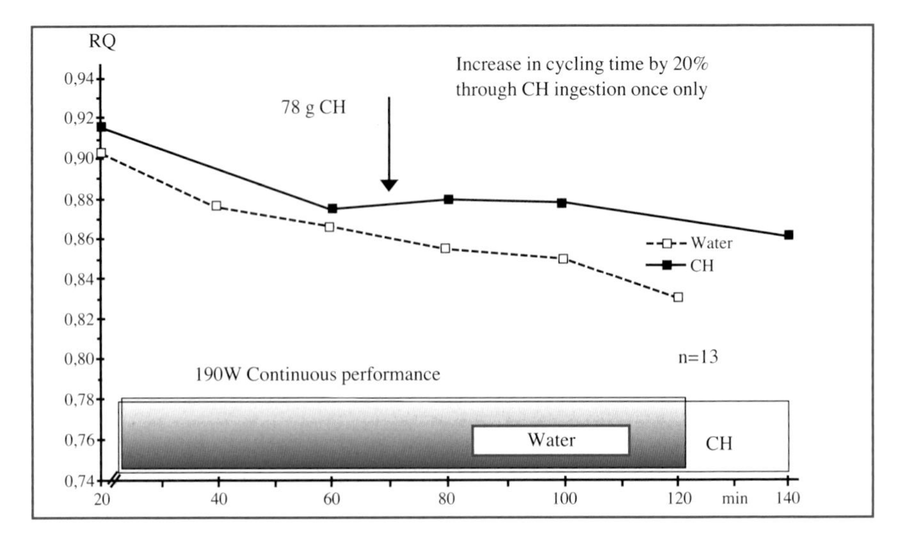

Fig. 18: Behaviour of the Respiratory Quotients (RQ: VCO_2/VO_2) during carbohydrate and water ingestion. When water is consumed, utilisation of free fatty acids is greater (assessed from the decrease in the RQ) than when 78 g of carbohydrates are consumed once. Also see Fig. 16.

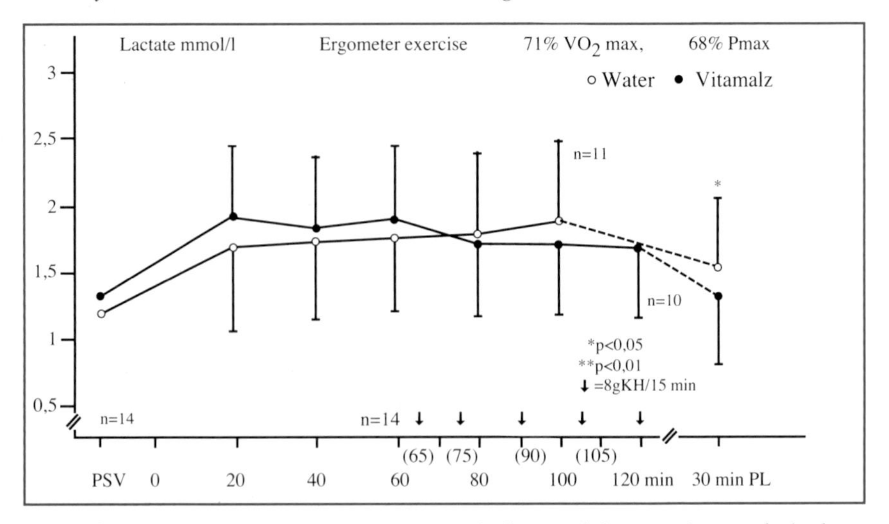

Fig. 19: Lactate concentration was not influenced by regular carbohydrate ingestion (Vitamalz ®) in quantities of 32 g/h during continuous exercise on the cycle ergometer in the area of the aerobic threshold.

3.3.3 Carbohydrate Ingestion After Exercise

The main aim of **immediate carbohydrate ingestion** after longer sporting **loads** is to fill up the emptied **glycogen depots** in the muscles and the liver. Especially in a state of reduced appetite glucose in any form must be consumed. For up to 60 min after exercise ends the prerequisites for the transformation of glucose to glycogen in metabolism are most favourable. The glycogen building enzyme **glycogen synthase** is most active after loading.

Consumption of concentrated **glucose solutions**, which, however, must be individually digestible, are advantageous for promoting regeneration. When one regains one's **appetite**, refilling of the glycogen stores should be continued through ingestion of natural carbohydrates. Rice, potatoes, bread and other carbohydrates (see Table 9) are suitable for this. Food with a carbohydrate emphasis after exercise has clear advantages over **mixed foods**. Through deliberate carbohydrate ingestion of about 6 g/kg of body weight even greatly emptied glycogen store can be refilled in 48 hours. Nevertheless, after exhausting running loads such as in a marathon or similar long duration loads, regeneration of the muscle glycogen depots takes 4-6 days, and of strength endurance potential about 8-10 days. This matches the experience that after greatly tiring load, willingness to run does not increase again until after the 4th day. **Glucose ingestion** is more suitable than **fructose** for refilling muscle glycogen. **Fructose ingestion** can be used with preference for building up **liver glycogen** again.

The quantity of carbohydrates to be consumed depends on the preceding energy consumption **(performance utilisation)**. After exercise with 3,000 kcal energy consumption it is higher than after those where only 1,200 kcal were utilised. If for example 2,000 kcal of energy were used up by loading, then 200 to 300 g of carbohydrates are necessary, corresponding to 3 to 5 g/kg of body weight. These quantities of carbohydrates result in 820-1,230 kcal, the difference from total energy consumption is covered by fatty acids and proteins.

In the first phase of regeneration after exercise, the **nutrient density** of the carbohydrates or their **glycaemic index** is decisive. The expression glycaemic index is used to express the effect of carbohydrates on the blood sugar level. Glucose and fructose have a considerably higher **glycaemic**

index than polysaccharides (e.g. starch in bananas), i.e. mono- and disaccharides lead most quickly to a rise in blood glucose concentration (see Table 10). In the interests of rapid regeneration the consumption of **fibre rich foods**, carriers of vitamins and minerals (e.g. vegetables, fruit) should be put off until **later**.

There are other measures for accelerating regeneration after sporting loads (Table 25).

Substances Aiding Regeneration

FUNCTIONS	SUBSTANCES
Energy Metabolism	Complex carbohydrates, creatine, branch chain amino acids (BCCA), medium chain fatty acids (MCT)
Micronutrients	Magnesium, zinc, selenium, chromium, Vit. C, Omega-6-fatty acids
Cell Protection	L-carnitine, Vit. E
Antioxidants	Vit. E, selenium, Vit. C, beta carotene, Vit. Q (Ubichinon)
Anticatabolics	Glutamine, branch chain amino acids, Hydroxymethylbutyrate (HMB), arginine, ornithine, carbohydrate-protein mixtures
Immune Stimulants	Red sunhat (echinacea), L-carnitine, artemisia abrotanum, mistletoe, camomile, arnica, salicylic acid, green tea

Table 25

3.3.4 Carbohydrate Ratio in Drink Solutions

Much is already known about the composition of beverages for athletes. There is no one beverage suitable for all situations occurring in sport, unless water is declared to be that beverage. Many factors influence **fluid ingestion**. These include the proportions of minerals (e.g. sodium chloride, bicarbonate, magnesium), the concentration of carbohydrates, consistency, temperature, the quantity drunk etc. Amongst the beverages for athletes those with a defined carbohydrate and electrolyte content have proved to aid performance. On the other hand, **electrolyte beverages for athletes** also contain additional glucose, long chain glucose units (glucose polymers), fructose, maltodextrines and other carbohydrates. The **combination** of carbohydrate beverages with minerals (electrolytes and trace elements) does not influence digestibility in the stomach or absorption. The prerequisite is that certain physiologically sensible concentrations of carbohydrate-electrolyte solutions are not exceeded (Fig. 20).

In the case of **loads in heat**, electrolyte solutions are preferable which only contain a **low proportion of carbohydrates**. Low proportions of carbohydrates in sports drinks stimulate water absorption in the intestines (BROUNS, 1993). This effect applies to solutions of up to 8%. In practice this means e.g. adding 80 g of glucose to one litre of mineral water. Even higher carbohydrate concentrations in fluids (over 10%) lead to reduced emptying of the stomach and stimulate the release of fluids from the intestine wall villi into the intestines for the purpose of equalising the concentration. The **osmotic gradient** determines the direction of the fluid current. Highly concentrated solutions must first be diluted before absorption in the intestines. This means a **temporary water loss** for the body. During exercise in cool surroundings, dilution of carbohydrates in the intestines has no major practical significance. For exercise in heat it is a different matter. Here ingestion of concentrated beverages in a state of dehydration can lead to a temporary additional fluid loss (see Table 5).

With regard to digestibility and absorption, concentrations of 6-8% (60-80 g carbohydrates/litre) are considered to be **optimal carbohydrate solutions**. If carbohydrates are consumed individually, this results in different digestible concentrations. The following **carbohydrate proportions** are recommended for **sports beverages** which are consumed during load, with regard to effectiveness, digestibilty and rapidity of absorption:

Fructose	up to 35g/l
Glucose	up to 80 g/l
Saccharose	up to 100 g/l
Starch solutions	up to 100 g/l
Maltose	up to 120 g/l*
Maltodextrine	up to 150 g/l*

** Because of their small molecule size, maltoses are still isotonic in higher concentrations, so that in practice even higher concentrations are digestible. With regenerational beverages the proportion of carbohydrates can always be higher.*

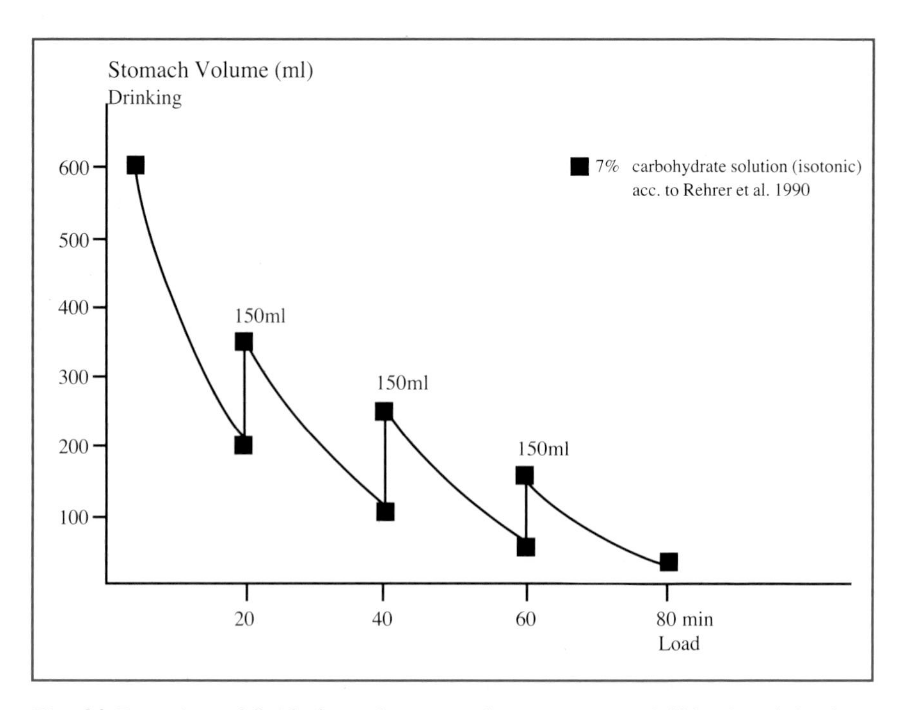

Fig. 20 Emptying of fluids from the stomach at ingestion of 600 ml and drinking 150 ml every 20 min. After REHRER and colleagues (1990)

Experience shows that for most athletes, and especially for **runners**, all **beverages** with concentrations of **up to 10%** of glucose, glucose polymers, saccharose, maltose or maltodextrine are digestible. For **fructose**, which is contained in many sports drinks, this **quantity is too high**. Fructose is added to the drinks because it does not influence insulin secretion. Here it is overlooked that glucose ingestion during long lasting loads only leads to slight insulin secretion which is not sufficient to cause a state of hypoglycaemia in endurance athletes (see Fig. 15). It has been proved that the opposite is the case. On consuming glucose or maltose solutions of 30 to 80 g/h there is a considerable rise in blood glucose concentration by 0.5-1.5 mmol/l (9-27 mg/dl). The increased blood glucose is a cause of the extension of performance and the improvement of muscle feeling on tiring.

In addition fructose ingestion has a further disadvantage. When beverages containing fructose, with concentrations over 3.5%, are drunk, some athletes have gastro-intestinal problems or diarrhoea. Furthermore the effect of fructose is delayed because it is first broken down into glucose in the liver and then transported to the needy muscles via the bloodstream.

3.3.5 Increasing Performance With Carbohydrate Ingestion?

Many active substances are assessed according to their effect in improving performance. On the basis of scientific criteria, in most cases no influence on performance capacity can be proved. **Genuine performance improvements** can only be achieved through regular **training**. Among the few natural substances which influence longer competition or training performances are **glucose** and other carbohydrates. The effect of glucose consumed during load is based on increasing or maintaining **blood sugar concentration** during long load duration. In order to secure the functioning of **cerebrum and cerebellum** during long duration exercise a normal blood sugar level is needed. If blood sugar concentration during endurance exercise goes down considerably below 3.5 mmol/l (63 ml/dl), the supply of glucose to the brain nerve cells is no longer maintained. **Consequences** of a **glucose deficit in the brain** are a loss of drive and motor disturbances, also called ataxia (Table 26).

Regular **carbohydrate ingestion** during long physical or sporting load thus has the effect of extending exercise duration and **increasing performance**. No competition, however, can be won just by increasing glucose ingestion. Taking glucose cannot replace training. After intensive exercise of over 90 min duration

Forms of Ataxia in Exhausted Athletes

- Coordination problems in locomotion
- Dysmetria (uneven stride)
- Equilibrium problems (staggering run)
- Consciousness restricted, athletes recognise motion failure and cannot influence it

Table 26

glucose ingestion prevents or delays hypoglycaemia. If carbohydrate solutions are drunk regularly and in appropriate quantities during long duration exercise, **less stress** occurs. The following muscle regeneration is more rapid because carbohydrate ingestion also reduces protein breakdown.

The state of the glycogen stores can be greatly influenced by nutrition at rest. In a state of hunger the **glycogen stores** are small, containing less than 0.6 g of glycogen/100 g muscle tissue. As a result of plentiful ingestion of carbohydrates during regeneration the glycogen stores in the muscles rise above normal: they can contain 2 to 4 g of glycogen/100 g of muscle tissue. BERGSTRÖM et al. (1967) recommended appropriate dietary measures for preparing for intensive endurance exercise which leads to an enlargement of the glycogen stores. These later became known more precisely as the **"Saltin diet"**. The basic idea is to achieve major emptying of the glycogen store during training before competitions by changing normal mixed nutrition to a **fat-protein diet** for three days. In the following three days very carbohydrate rich food is eaten to create **glycogen over compensation** in the muscles. With glycogen over-compensation 3 to 4 g of glycogen per 100 g muscle tissue is stored. This is twice the normal amount. The full glycogen stores have a positive effect on performance capacity, especially in all forms of intensive and longer endurance exercise.

As **failures** were more common than successes with these dietary measures, they are today seldom applied, or only in individual cases in top level training where a great deal of experience is available. For the preparation of personal top performances, adjusting training load, i.e. **load reduction** before important competitions, with **simultaneous ingestion of carbohydrate rich foods**, has proved more effective than carbohydrate withdrawal variants over several days and training breaks.

Physical Peculiarities at Medium Altitudes

Gravity:
Decrease in earth acceleration per 1,000 m by 0.3 cm x s^{-2}
(Weight relief at 2,000-3,000 m altitude under 0.1%)

Air Resistance:
Exponential decrease in air density with increasing altitude
(Air density decrease at 1,800 m 20%, 2,500 m 26% and 3,000 m 31%)

Energy relief in 5,000 m run at 3,000 m altitude
(about 3.4%, in cycling 28%)

Temperature:
Decrease in temperature of 1°C per 150 m altitude
(Risk of hypothermia during training)

Radiation:
Increased UV radiation and cosmic radiation
(UV radiation at 295 nm increases by 35% per 1,000 m)

0_2 Partial Pressure:
Parallel with air pressure, 0_2 partial pressure decreases exponentially
(at 3,000 m 1 l of air contains 31% less oxygen; at 2,200 m 24%)

Table 27

3.3.6 Protein Ingestion During Exercise

In the case of long duration endurance performances there is not only major **glycogen depletion** but also **contra-regulation** in metabolism, which consists of redevelopment of glucose in the liver **(gluconeogenesis)**. This is necessary to maintain blood sugar concentration during load. **Gluconeogenesis** results from glycerol, lactate and **amino acids**. During long duration exercise mainly amino acids are broken down to create new glucose. The amino acid store is not very large; it is only about 110 g. The proteins involved in functional exchange and immunological defence are not part of this direct reserve for energy consumption. They can, however, be used in emergency situations. Greatly loaded athletes often have concentrations of immunoglobulins (Ig) at the lower norm limit, especially IgG.

The amino acids mainly involved in gluconeogenesis are the **branch chain amino acids** (valine, leucine and isoleucine) and **alanine**. In metabolism about 0.6 g of glucose come from 1 g of amino acids. Meanwhile it is known that in addition to the branch chain amino acids, during exercise others also serve as precursors for gluconeogenesis. In addition to alanine this applies to **glutamine**. Up to 10% of **energy requirements** can be covered by **oxidation** of the **amino acids** during long duration exercise. For example, in a marathon 30 g and in a 100 km run 90 g, of amino acids are broken down.

The greater the quantity of amino acids broken down, the longer regeneration of performance capacity takes. After a 10 km competition run an athlete recovers more quickly than after a marathon. Recent research has shown that ingestion of branch chain amino acids during a long duration exercise or during altitude training led to an improvement in performance (PARY-BILLINGS et al. 1992; BIGARD et al. 1993).

4 Environmental Influences and Nutrition

4.1 Altitude Training

In contrast to low altitudes, medium altitudes have a number of peculiarities which have an influence on training (Table 27).

More and more often altitude training is becoming a component of competitive training, especially in endurance sports. Successful middle distance runners live at medium altitudes. At the **usual training altitudes** of 1,700 m to 3,700 m in competitive sport, great significance is attributed to securing load tolerance, fluid balance and a carbohydrate and protein rich diet. The **increased carbohydrate requirements** in altitude training are connected with the adjustment of metabolism towards **mainly carbohydrate combustion**. The higher oxygen content in the basic structure of carbohydrates, as opposed to low-oxygen fatty acids, leads to increased carbohydrate combustion when there is an oxygen deficit. At the accustomed load level there is an increase in lactate concentration of 1 to 3 mmol/l. Under accustomed training loads the increased carbohydrate combustion and glycolysis lead to **early exhaustion** of the glyocogen stores. To take the pressure off the glycogen stores, at altitudes of 2,300-3,200 m training is at a considerably **lower intensity**. Of the sports practised at higher altitudes, the greatest reduction of load takes place in running. In aerobic basic training the reduction is 5-10%. Repeated altitude training seems to be most effective; in successful endurance sports altitude training is repeated 4-6 times a year (altitude chains).

In altitude training, chronic glycogen deficiency leads to additional **protein breakdown**. The increased protein breakdown can be measured using the increase in serum urea concentration. The increased protein catabolic state during muscle use under an oxygen deficit leads to constantly increasing **residual fatigue**. The increase in protein use and protein breakdown forces athletes to take two key measures in altitude training, namely lengthening rest intervals and adjusting diet. Despite reduced load intensity, in altitude training rest intervals must be longer. In nutrition, increased attention must be paid both to the quantitative and the qualitative aspects. The result is deliberate and plentiful consumption of carbohydrate and protein rich foods. It should be noted that when an accustomed diet of whole foods is kept to during altitude training there is increased production of intestinal gas (meteorism). This can be disturbing during group training. In order to provide a balance, vegetarian oriented athletes should take products to altitude training which contain high valency proteins (amino acids).

At high altitudes, increased fluid loss via the respiratory tract is not electrolyte or mineral loss. Only if sweating is heavy is it necessary to add minerals when drinking.

The **risks** of altitude training are thus **glycogen deficiency**, increased protein breakdown and major **dehydration**. Together with the greater effect of load stimuli these lead to **longer regeneration time**. In altitude training recovery times are longer than at normal levels. **Improved performance** does not come directly after altitude training. Only after a transformational period of 2-3 weeks at normal altitude does the performance capacity of most athletes improve. As a result of volume training reduced by 20-30% at normal altitude there is usually an improvement in performance on the 14th to 17th day. The body needs this time to process the stimuli. An unsuccessful first altitude training should be analysed thoroughly before jumping to conclusions about its usefulness. One of the causes may be a diet not best suited to the conditions.

4.2 Training in the Cold

By wearing warm and special sport clothing, it is possible to train in winter sports for several hours at outside temperatures of up to -20°C. In cross-country skiing these temperatures are the limit permitted for starts in competitions. In running, training in the cold is done at low velocities. Because of the risk of hypothermia of the feet, training in cycling is limited timewise and is not possible for several hours at a time except with special leggings as well as hand and face protection. Endurance training at minus temperatures is linked with considerably **increased energy utilisation**. It takes a few days to raise basic utilisation when in the cold; initially athletes feel an increased need for warmth when exposed to the cold and are constantly cold. Increased energy utilisation calls for ingestion of high calorie foods. This requirement can be easily met just by increasing the **fat ratio** in the diet. Winter training increases energy requirements by about 10%; to protect against the cold the body develops an insulating layer of subcutaneous fat.

Swimming for longer periods in cold water (from 15°C to 20°C), as is common e.g. in triathlon with a **wetsuit**, also increases energy requirements. **Long distance swimmers** who aim to reach distant destinations across lakes or seas ("Channel swimmers") are also a special case. To secure their body core temperature when swimming in water that is 15 to 18°C "warm" they need to "feed up" a thicker layer of subcutaneous fat, otherwise they have no chance and have to give up owing to hypothermia. An over-caloried diet is obligatory for long distance swimmers. These athletes are usually very heavy (80 to 100 kg). During hours of swimming, complex carbohydrates in warm fluids must be consumed (700 to 800 kcal/h). Food consumption in the water must be plentiful as hypoglycaemia would be fatal!

In winter training, no matter what sport, no compromises need to be made to one's usual diet. Increased appetite due to exercise under cold needs energy ingestion. Because of only light sweating disturbances to the mineral balance rarely occur in winter training. **Ingestion of vitamins** becomes very important, especially when the local supply of fruit and vegetables is limited. In winter special attention should be given to vitamin C intake.

4.3 Training in Heat

Training in heat, i.e. at outside temperatures above 25°C, is usually linked with visits to southern countries, different time zones or indoor training. In central Europe heat training only takes place for short periods; there is no acclimatisation. During brief heat episodes in Europe there is adjustment regulation, for which the body needs one or two days. By moving training to the early morning or late evening, heat can be avoided. This applies especially to endurance training. If starts during heat are planned in other countries, previous training should be at the warmest time of day, supported by warm sport clothing. The prerequisite for heat acclimatisation is active daily training of 2-4 hours for ten days at temperatures above 30°C. If a short journey to other countries allows no time for acclimatisation, **air-conditioned rooms** should be avoided before important starts. At rest or indoors, unnecessary sweating should be avoided because it increases mineral loss.

In great heat food consumption must be adapted to the changed conditions. **Easily digestible carbohydrates** should be given preference.

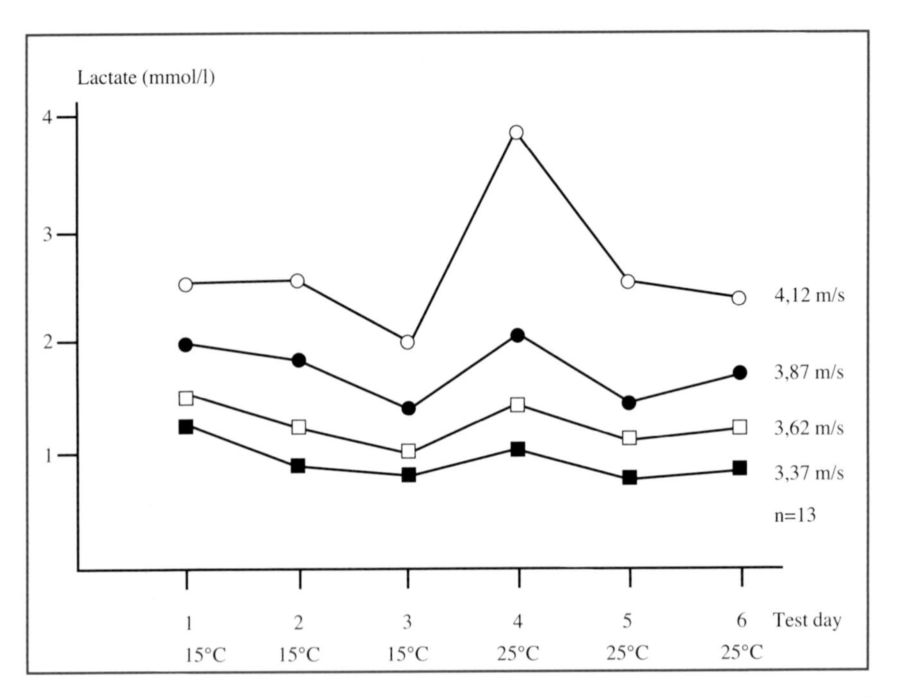

Fig. 21: Behaviour of lactate concentration during 4 x 2 km graded exercise on the treadmill with varying speeds (m/s). On the 4th day room temperature is suddenly raised by 10°C (from 15 to 25°C). At higher speeds there is a considerable rise in lactate which was not evident during the exercise following.

When training in heat, energy consumption rises. Biological effort increases under heat load. In **training control** the greater effort can be seen in the higher heart rate and/or lactate concentration. At comparable performance lactate concentration increases by 1-2 mmol/l.This metabolic adjustment is an expression of the added involvement of **glycolysis** to secure performance (Fig. 21). In heat training speed should consciously be reduced. Maintaining accustomed training performance leads to premature raising of the body core temperature and increases load stress. In endurance competitions in heat it is advantageous to **build up slowly in the first sector**; this always pays off in the closing phase of the race. Most athletes who begin competitions too quickly experience a great drop in performance capacity towards the end of the load. The drop in speed is much higher than usual under heat loads.

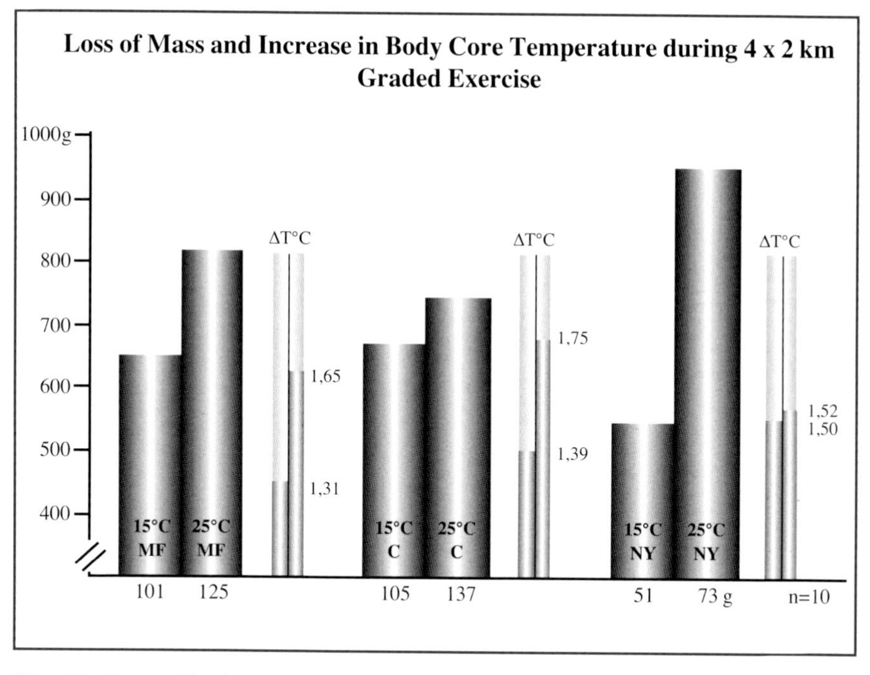

Fig. 22: Loss of body mass after exercise of 4 x 2 km (8 km) on the treadmill at 15°C room temperature. Nylon (NY) clothing caused the greatest release of sweat and also cooling (least rise in body core temperature: hatched narrow column). Cotton (C) took a middle placing in sweat loss at 25°C in comparison to microfibres (MF) and NY.

The skin and especially the head should be protected from unaccustomed UV rays with suitable **headwear**. Wearing a white peaked cap and sports sunglasses with UV glass is advantageous. If the sun shines from behind, additional neck protection should be worn. In strong sunshine, sun block (**light protection factor** above 10) must be applied to unprotected skin (nose, ears, shoulders, mouth area). Increased **ozone concentration** in summer has turned out to be a new disruptive factor which causes irritations in the respiratory tract of runners in particular. The ozone concentration which irritates the respiratory tract is about 200 μg/m3. The first to be affected are untrained persons with asthma and athletes at high running speed. At 1,500 μg/m3 (0.75 ppm) of ozone, oxygen intake of trained athletes has been shown to be reduced (WILMORE/ COSTILL, 1988). The level of ozone concentration depends on the time of day; it is highest between 11 a.m. and 5 p.m. in areas with particularly "clean air" and bright sunshine.

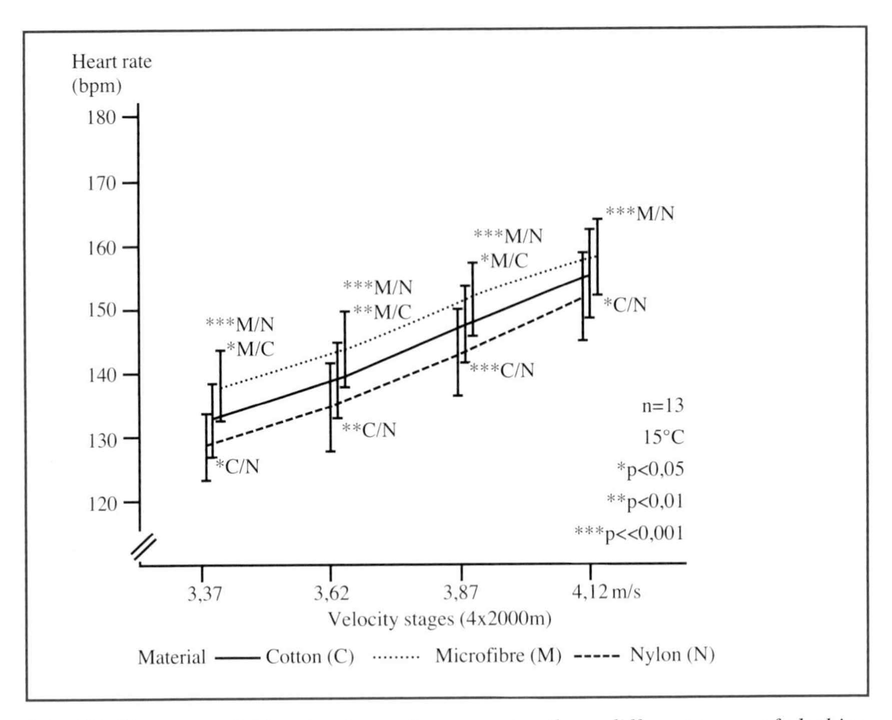

*Fig. 23: Heart rate (Hr) behaviour when wearing three different types of clothing at a room temperature of 15°C. Microfibre warmed the most and nylon cooled best. Cotton took a middle placing in the influence on Hr. Significant differences are given in *.*

Training should therefore be moved to the early morning and/or late evening hours when ozone levels are high. Outside Europe (e.g. Australia) a considerably higher ozone concentration which impedes breathing during training must be reckoned with.

In heat training, **fluid consumption** must be greatly increased, paying attention to a plentiful electrolyte content. **Diet** should be adjusted to a high **carbohydrate ratio** and **vitamin content**.

In **heat competitions** effective prevention lies in constant and early consumption of plenty of fluids containing electrolytes. Drinking must commence before eating. One should begin drinking after the first 15-20 min of exercise and continue every 15 min. As a precautionary measure it is possible to drink before exercise starts, whereby the quantity of fluids consumed must not hinder starting tempo. **Heat load** can also occur in sports halls. One's behaviour should then be analogous to outside. Indoors, however, there is no cooling effect of wind and no sunshine.

Drinking greatly cooled beverages leads to gastro-intestinal problems and training hindrances. This also applies to competitions where with good intentions athletes are given deep chilled drinks, which often results in **stomach cramps**.

In **tropical countries**, for health preventive reasons, no tap water should be drunk. Uncooked food in general should be avoided. This applies especially to raw salads, ice cream, fruit and unpasteurised milk.

With **sports clothing** only limited heat protection can be achieved. The **best cooling effect** during short loads is provided by nylon because it leads to the greatest sweat production (Fig 22). **Cotton** soaks up much sweat and takes a middle placing between nylon and microfibre with regard to its cooling effect. Available findings allow assessment of the cooling effect of cotton using its effect on heart rate (Fig. 23).

In comparison to cotton and nylon, sports clothing made of **microfibre** warms relatively well and should be avoided during longer exercise in heat despite its good drawing off of sweat.

4.4 Changing Time Zones

As a result of competitions, altitude training camps or climate camps, a changing of time zones occurs increasingly often. The **biorhythm**, which has adjusted to local conditions of the day-night rhythm, is disturbed on arrival in a new time zone. The light-dark centre in the optic nerve junction (chiasma opticum) with its special receptors probably plays a key role in the adjustment process. Recent findings have shown that melatonin released by the pineal body has a steering effect on the light-dark rhythm and the waking state (REITER/ROBINSON, 1997).

The **biorhythm**, settled at a particular local time, remains adjusted to it when the location changes; this applies to the sleeping rhythm, body core temperature, the feeling for hunger, the rhythm of excretion of urine and stool, the immune system among others.

In order to quickly adjust to a new time zone, immediately after arrival physical exercise or a light **training** should be begun. **Food consumption** also has to be adapted to the new local time. In the first days eat more frequently, giving preference to light, easily digestible foods. These also include energy bars taken along. It is advisable to take accustomed foods to countries which have considerably different eating habits. Contact with unaccustomed germs in food (e.g. eschericha coli) which are harmless to the locals can take on dangerous proportions for athletes. The main adjustment problems overseas affect the gastro-intestinal tract. Diarrhoea should be taken seriously right from the start and training should be interrupted immediately so that fluid loss does not increase.

The **central nervous system** has the longest duration of constancy in its **biorhythm** and needs one day per hour of time difference to adjust. Healthy people and athletes can handle a two hour time change without problems. In the case of **intercontinental flights** with 6-8 hour **time differences** a minimum of 3-4 days of adjustment in the main functions is to be expected. Top athletes too must allow one day of adjustment for each hour of time difference. Top athletes experience their greatest drops in performance when they ignore the necessary time adjustment and heat acclimatisation. In order to achieve top sporting performances on other continents, sufficient adjustment time must always be adhered to. For athletes with less training experience the adjustment probably takes longer than for those with more training. Flights from **east to west** are dealt with best because the waking state is extended. So-called **"jetlag"** is not so great as with flights against the course of the sun, i.e. from west to east. Intercontinental flights from west to east, e.g. to Australia via Asia, need to be undertaken at least ten days in advance for a sporting peak.

Taking **melatonin** in a dosage of 5 to 10 mg can considerably reduce the effects of jetlag. At the destination it is recommended to take 5 mg of melatonin one hour before sleeping. This makes it easier to get to sleep, improves the quality of sleep and resets the "internal clock"(LINO et al. 1993). Currently melatonin is only available as a dietary food in the USA. In Germany, **Diazepam** has proved helpful in aiding sleep and relaxation. Longer **north-south flights** and vice versa have no influence on biorhythm. These changes of location require few changes in diet (e.g. South Africa). The increased development of intestinal gas (meteorism) during flights is due to reduced air pressure in the cabin which usually occurs at an altitude of 2,000 m.

Immediately after arriving in a new time zone **top class athletes** should try to adjust **food intake** to the **local daily rhythm**. If their condition allows, they should begin training. Full sporting perfomance capacity only returns when the adjustment in the sleeping-waking rhythm is completed, especially in the case of intensive training. If an athlete lacks motor drive for training load because of great fatigue and exhaustion, he should first get enough sleep and then train. Perhaps the approval of melatonin in Germany will ease this situation, for French athletes have had good results overcoming jetlag problems by taking 8 mg of melatonin when returning from the USA (CLAUSTRAT et al. 1992).

5 Fluid Intake in Sport

For a long time there were varying views on fluid intake during sporting loads. Among athletes the **false idea** that one should not drink during long lasting exercise was stubbornly held onto. Views that fluid intake during marathons should be avoided were particularly extreme. **Abstinence from drinking** was seen as strengthening the will to perform and it was assumed that dehydration had no influence on performance capacity. These assumptions in practice have proved to be fundamental errors. Meanwhile numerous studies have shown that **fluid intake** is necessary before, during and after exercise and clearly has the effect of **aiding performance**.

The advantages of drinking during long lasting exercise are derived from the physiological finding that during exercise sufficient fluid must always be available to secure the functions of the **cardiovascular system** and the **heat balance**. Dehydration of the body caused by fluid deficiency increases the heart rate (Hr) during exercise and accelerates the increasing of body core temperature. On the other hand, sufficient fluid intake has the effect, especially during exercise, that the regulation of the Hr to a normal level, and the increase of body core temperature occur more slowly. Regular fluid intake delays the rising of body core temperature and prevents **overheating (hyperthermia)**.

5.1 Fluid Intake and Performance Capacity

Sufficient fluid intake during exercise is often also neglected by experienced athletes. The **feeling of thirst** as a regulator of fluid intake is not always reliable. Often **dehydration** reaches over 3% of body weight before an athlete registers this state in the form of feeling very thirsty (see Table 5). In a state of dehydration fluid intake has priority over consumption of food. In cases of major dehydration there is a great decrease in locomotion velocity. Additionally, after all long lasting loads the glycogen depots are practically exhausted and performance is mainly maintained by fat metabolism.

Storage of water in the body is limited; excess water is excreted via the kidneys.

During exercise fluid loss is mainly via sweat (see chapter 5.2). Adults can handle the loss of 1-3 l of sweat during a long lasting load without a major drop in performance (see Table 5). Because **absolute fluid loss depends on body mass**, this is expressed as a percentage of body weight. The loss of a litre of sweat in a child weighing 35 kg results in greater disruption than in an adult weighing 75 kg. In this example the child's fluid loss would be 2.8% while for the adult it would only be 1.3% of body weight.

Under normal outside temperatures **fluid loss** is dependent on load intensity and in particular on load duration. The **sweating rate** is limited, namely 1 to 1.5 l/h. During heat loads in endurance sports, however, the sweating rate can be over 2 l/h. Serious disruptions to the fluid balance are only to be expected with exercise of over 60 min duration. The exception is great heat (over 30°C), combined with high humidity. This combination causes extreme fluid losses even in a very short time. With **heat loads**, in comparison to exercise at normal outside temperatures (12 to 23°C), disruptions to the water and electrolyte balance occur earlier. Additionally attention must be paid to the effect of combinations of outside temperature and wind conditions on fluid losses. Here coaches must constantly remind athletes to drink plentifully. Changes in the outside temperature (increase), the dynamics of air humidity and wind conditions during long lasting competitions should always be taken into consideration and calculated in advance. In addition to fluid and food consumption they also influence the choice of **sports clothing**; this must be selected according to the expected rise in temperature and not the coolness at the time of the start. If an athlete has dressed too warmly because of the cool temperature at the starting time (which usually happens in the morning when one is tired), he should be prepared for a change of clothing. Through unnecessary sweating one can organise performance prevention oneself. For example, analyses over many years showed that in the Boston Marathon, relatively the best running times were recorded at an outside temperature of 8 to 12°C. Of course clothing depends on the running speed (performance capacity) to be expected. In city marathons in cool seasons it can repeatedly be observed that the first hundred runners wear shorts and the following slower runners more and more often have long trousers.

Drinking tap water during sporting exertion is **not recommended** because it contains too little salt and hardly any other minerals. For a long time American scientists advised drinking normal water during runs in heat. Because of its low osmotic pressure (low molecular density) water seemed suitable for the prevention of major dehydration. More recent research of this problem, however, has shown that consumption of great quantities of low-mineral tap water can lead to **"water poisoning"** in the body.

Disruption of the fluid balance through drinking too much water was first described in 1985 by South African researchers (NOAKES, 1992) and later by participants in the Hawaii long triathlon (at 30°C air temperature). Those affected were people who competed at a relatively slow tempo and drank a great deal of tap water at an outside temperature of 30°C. The replies on quantities drunk ranged from 10-15 l during the competition, which lasted over ten hours. The collecting of the low sodium water in the intestines leads to migration of sodium from the body into the intestines. This **reverse electrolyte compensation** occurs because water can only be absorbed when a certain sodium content is present. Only the salt released by the intestinal villi makes it possible to absorb the water. Because in addition to this salt loss much salt is lost through sweat, a major decrease in the blood sodium concentration occurs. The decrease in blood sodium concentration to below 130 mmol/l is the cause of **brain function disturbances** with restrictions of consciousness and major coordination problems (see Table 26).

During endurance loads of 3-4 hours duration under European climatic conditions (at 15°C to 25°C outside temperature), on the other hand, no problems with the sodium balance are to be expected. Owing to their diet and limited sweating during training, athletes have **sufficient salt reserves** even for one-off extreme situations. Many foodstuffs contain hidden salt for preservation purposes. Thus at a level of salt consumption of 10-15 g daily competitive athletes have no need to consume additional salt during competition or training.

A slow drop in sodium concentration in the blood to just below 130 mmol/l rarely leads to problems. Rapid salt losses, or decreases in sodium concentration in the blood, are more dangerous. In particular, a rapid drop in sodium concentration from 140 to 125 mmol/l in a short time can lead to disruptions to the **brain function**.

When there is a **sodium deficit** athletes feel **great fatigue**, lack of concentration, loss of drive and daze. In **extreme cases** of salt deficiency central **cramps** and **unconsciousness** occur. These disruptions are caused by water flowing into the brain, a **brain oedema** occurs which is expressed in the symptoms described. The sodium deficit in the blood disrupts the osmotic equilibrium at the blood-brain water barrier. Collapses of athletes in heat must be taken seriously and require medical help.

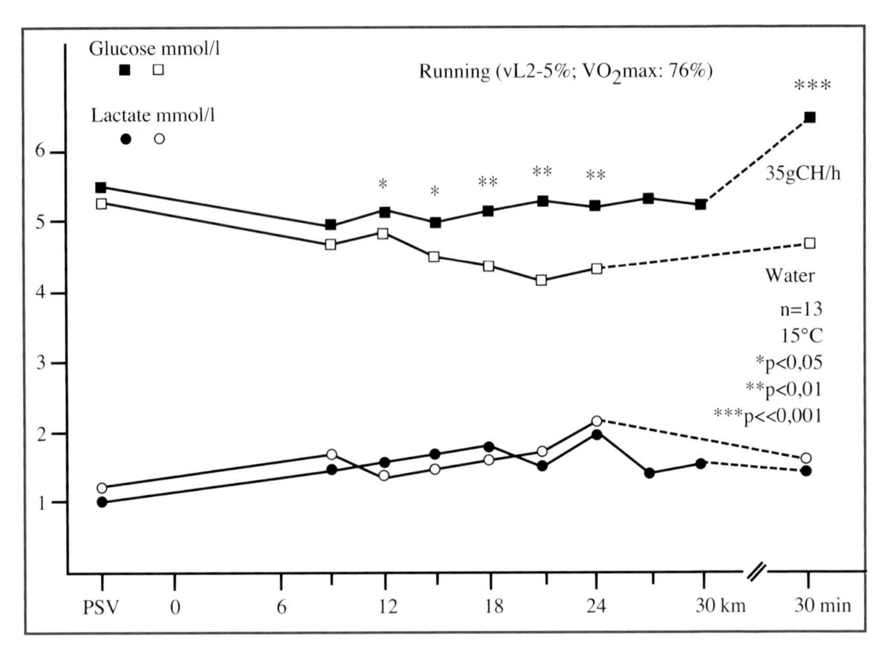

Fig. 24: Behaviour of glucose and lactate during continuous exercise on the treadmill at an intensity of 5% below the speed of the aerobic threshold at Lactate 2 mmol/l (vL2). Glucose consumption increased blood glucose concentration but not lactate concentration.
*Significant differences are shown in *.*

Fluid requirements for exercise of several hours can be calculated in advance (see chapter 5.5). This is not so when sodium enriched water (mineral water) or glucose solutions are consumed. Electrolyte or glucose solutions secure longer lasting performance capacity in comparison to drinking water only (Fig. 24). **Carbohydrate-electrolyte beverages** are an **ideal fluid replacement** for endurance athletes during and after heat loads.

The number of molecules dissolved in a fluid determine molecular pressure (osmolarity). All solutions corresponding to an average molecular pressure in the body of about 300 mosmol/kg are called isotonic (Table 28). **Isotonic beverages** are those with a molecular density of up to 330 mosmol/kg. This is similar to a 0.9% salt solution, which is isotonic. If solutions have more soluble molecules they are considered **hypertonic** (> 500 mosmol/kg). Hypertonic solutions occur when fruit juices are consumed. This is the reason for the recommendation to dilute these with mineral water (e.g. apple juice-mineral water mix at a **dilution** of 1:1 to 1:3.). Dilution of 1:2 means e.g. one part fruit juice to two parts water. Dilution is also a good idea because fruit juices have a high **potassium content**. The potassium content is 1,000 mg/l on average. Thus drinking pure fruit juice during loads is inadvisable. When diluting juices, care should be taken that the amount of potassium in the mixed drink does not go above 220 mg/l. This quantity of potassium compensates the loss of potassium in sweat which contains about 225 mg/l of potassium during exercise. Too high a potassium concentration could have a negative effect on the increased blood lactate concentration.

Fluids with very few soluble molecules are called **hypotonic** (< 200 mosmol/kg); the best known hypotonic fluid is **tap water**. In its extreme form it is distilled water arising from melting snow.

Many **rehydration beverages** are recommended for athletes. REUSS (1992) and BROUNS/KOVACS (1996) suggested a solution according to physiological findings and adapted to conditions (see Table 29). A number of factors which have an influence on the absorption of fluids and nutrients need to be considered in a state of dehydration (see Table 28).

Factors Influencing Absorption of Fluids and Nutrients during Endurance Exercise

1. **Fluid amount**
   (optimal 100 - 150 ml/serving)
2. **Energy content**
   (5 - 10% carbohydrate solutions preferable)
3. **Osmolality**
   (≈ 300 mosmol/l is isotonic. Dilute fluids with higher osmolality with water)
4. **Load intensity**
   (Loads of over 75% of VO_2max reduce digestion; not fluid consumption)
5. **Load stress**
   (Psychophysical stress and fear reduce digestion)
6. **Dehydration**
   (Digestion slowed)

Table 28

Requirements of an Optimal Beverage to Aid Rehydration and Regeneration*

Carbohydrates	60-80 g/l
Sodium	1,4-1,0 g/l
Chloride	< 1,5 g/l
Potassium	< 225 mg/l
Calcium	< 225 mg/l
Magnesium	< 100 mg/l
Osmolarity	200-300 mosmol/l
Acidity (pH)	> 4,0

Univeral Mineral Drink for Endurance Athletes**

Sodium (Competitive athletes)	30-40 mmol/l
(Fitness athletes)	10 mmol/l
Potassium	5-6 mmol/l
Calcium	1 mmol/l
Magnesium	2-6 mmol/l
Iron (with vit. C)	6 mg/l
Zinc	3 mmol/l
Copper	1 mg/l
Carbohydrates (of which 50% glucose)	20-40 g/l
Osmolarity (isotonic)	250-300 mosmol/l

Table 29:

* *Amounts according to BOUNS and KOVACS (1996).*
These requirements are met e.g. by Isostar ® and Gatorade ®.

** *Amounts according to REUSS, 1992*

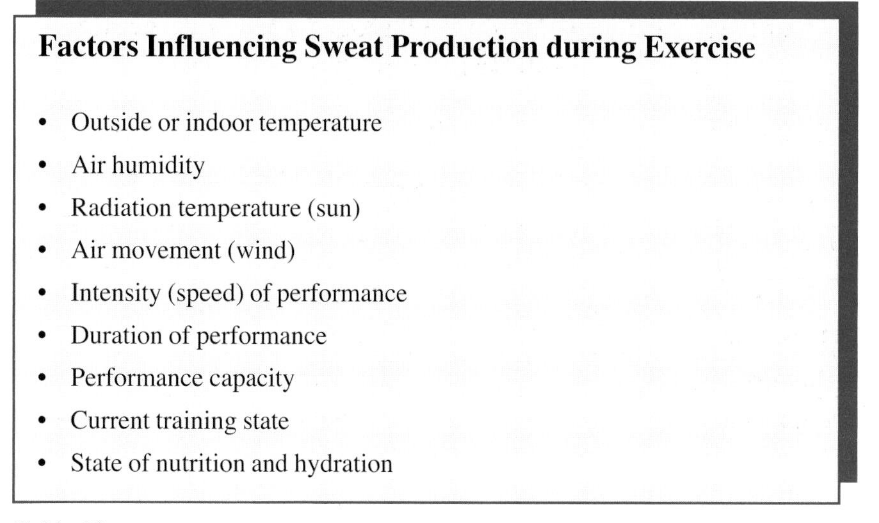

Factors Influencing Sweat Production during Exercise

- Outside or indoor temperature
- Air humidity
- Radiation temperature (sun)
- Air movement (wind)
- Intensity (speed) of performance
- Duration of performance
- Performance capacity
- Current training state
- State of nutrition and hydration

Table 30

Consequences of Sweat Loss

Sweat production is influenced by a number of factors (Table 30).

The first consequences are a **decrease in plasma volume** (hypovolaemia) and a relative increase in the number of blood corpuscles (erythrocytes, leucocytes). This **haemoconcentration** can be ascertained from the decrease of the haematocrit. The **haematocrit** expresses the relationship between the solid and fluid components of the blood. Endurance trained athletes have a haematocrit of 0.4 to 0.44. If the haematocrit rises above 0.55 (55%) it results in a hindering of blood flow in the capillaries. In this case the solid components dominate and the exchange of gases in the capillaries is hindered. These **micro-circulation** disruptions also make the exchange of energy substrates more difficult so that performance capacity can decrease as a result. For example, in a state of **hypovolaemia** (increased haematocrit) lactate remains in the muscles longer, and transport to the liver and other organs to be broken down is delayed. Further serious consequences of **dehydration** are disruptions to heat exchange. If there is also muscular fatigue as well as electrolyte problems, **muscle cramps** can occur. Generally muscle cramps result from a local disruption of the supply of magnesium and/or calcium. Electrolyte disruptions alone, however, do not explain the occurence of muscle cramps which usually occur in the last third of long duration exercise (e.g. marathons).

Degrees of Heat Exhaustion in Sport at High Outside Temperatures

(according to NEUMANN and SCHÜLER, 1994)

Cause:
Major water and electrolyte loss during sporting loads.
Rise of body core temperature to 41°C. Disruptions to central nervous system.

Symptom: Slight heat exhaustion:
Sweat loss of about 1.5 l (2% of body weight). Rise of body core temperature to 39-40°C. Considerable sweat production. Indications of circulatory centralisation ("goose bumps"). Great thirst and tiredness.

Major heat exhaustion:
Sweat loss about 2-4 l (3-6% of body weight). Athlete breaks off load. Rise of body core temperature to 40-40.5°C. Considerable sweat production. Indications of major agitation of the central nervous system (headache, locomotor unrest).

Very serious heat exhaustion:
Sweat loss above 4 l (6-8% of body weight). Sudden breaking off of load. Collapse. Increase of body core temperature above 41°C. Failure of sweat release (dry skin). Disruption of nerval regulation (nausea, headache, apathy). Beginning dimming of consciousness (disorientation).

First Aid:
Lie flat, immediately begin cooling measures. Measure body core temperature. If consciousness is retained, give fluids and glucose. In serious cases call a doctor or take to hospital. Do not underestimate consequences if condition seems to improve; always arrange for a doctor to see the patient.

Table 31

As a result of disrupted **oxygen intake** in muscle tissue, reduced **energy exchange** and increased **body core temperature** above 40°C, in a state of dehydration, a decrease in performance capacity occurs. Visible consequences of the disruption to the water and electrolyte balance are abrupt; performance drops which soon after can lead via locomotor steering errors to a restriction of consciousness (see Table 26). Such serious control derailings during dehydration can only be prevented by early consumption of mineral enriched fluids, which should also contain glucose. The key preventive measure for avoiding hyperthermia, however, is a good training state and well dosed fluid consumption (electrolytes). Under heat loads personal best performances should not be attempted.

The effectiveness of fluid consumption depends on the solution's osmotic pressure. This is created by the mineral and the glucose content. As already mentioned, **tap water is not suited to fluid compensation in case of dehydration** because the proportion of soluble minerals is too low. If only tap water is drunk, problems in the gastro-intestinal tract are pre-programmed and thirst cannot be quenched.

Optimum fluid replacement in heat should always be oriented to electrolyte and glucose solutions. Consumption of electrolyte mixtures should be complemented with a proportion of 4 to 10% glucose or maltodextrine solutions. Enriching sports beverages with these quantities of carbohydrates, and possibly also small amounts of salt, does not delay emptying of the stomach and absorption in the intestines in comparison to water (BROUNS et al., 1991). This procedure is based on the experience that dehydration during exercise of over 60 min duration are often combined with hypoglycaemia. The proportion of carbohydrates in the drink ensures the blood sugar level is maintained and contributes to extending load intensity.

5.2 Fluid Intake and Temperature Regulation

Sporting performance of up to an hour duration is possible in heat without major drops in performance. An effective compensation system works through s**weat production** and **evaporation** on the skin. 70 to 80% of the heat created by muscle work can be released through sweat evaporation. If insufficient heat is drawn off, the body core temperature rises. In the case of strenuous sporting loads, body core temperature of 39 to 40°C are the rule. Body core temperature increases above 40°C, however, are rare, they occur during marathons in heat. If the body's own cooling mechanisms are not enough during longer lasting exercise, and the athlete neglects to drink plentifully, there can be a rise in body core temperature to over 41°C. That leads to various forms of heat exhaustion of the athlete (Table 31).

The most effective form of **prevention of hyperthermia** of the body is to **drink fluids early and plentifully**. On average **1l of fluid per hour of exercise** should be consumed. This amount should be divided into 3-4 servings, i.e. every quarter of an hour 250 ml should be drunk. When running, fluid consumption that is too high causes high pressure on the stomach. By drinking in mouthfuls this state in sensitive athletes can be avoided. There are major individual variations in the digestibility of fluids at varying temperatures. Cool fluids (5 to 8°C) pass through the stomach more quickly than warm liquids.

During a marathon in Japan the athletes' own drinks were cooled so much that they had stomach problems.

An ample supply of fluids during exercise leads to slower increase in body core temperature and thus reduces the danger of heat damage. Unexpected drops in performance are known during runs in heat; they can be avoided to a great degree by drinking regularly. By adding **0.5 to 1 g of salt per litre** in self-made water solutions these can be absorbed more easily, and water is not drawn unnecessarily from the body.

5.3 Fluid Intake in Sport Groups

a) Endurance Sports

In endurance sports **fluid intake** is dependent on **exercise duration**. Sporting loads of up to 60 min duration require no additional fluid intake in heat. A fluid loss of 0.5 to 1l does not affect performance capacity. At running events in heat of over 70-90 min duration a strong need to drink can be expected among less well trained athletes. This real situation must be planned for both personally and by the promoter. In this situation a plentiful supply of drinks (carbohydrate-electrolyte) must be arranged. Fluid requirements are twice as high as usual.

As already mentioned, in **endurance competitions** (cycling, long distance running, triathlon, cross-country skiing etc.) fluid requirements of one litre per hour of exercise are planned for. The basis of this quantity is the average performance capacity of the athletes, and the outside temperatures up to 23°C with a slight wind and low humidity.

If the outside temperature during the race rises above 30°C, water and drinking fluid requirements increase drastically. Water is also used to cool the body. From an athlete's appearance the **rate of sweat** production can be estimated. If sweat production is only about 0.5 l/h, sweat is not visible on the skin. Clearly **visible sweat** indicates sweating rates of about one litre/h. **Dripping sweat** is a sign of maximum sweat production (about 1.5l/h).

Fitter athletes sweat less than athletes who are not so fit at the same level of performance. **Well trained athletes drink less than averagely trained athletes**. This behaviour can be observed regularly in training and in competition. A number of studies of **drinking behaviour** in marathons at 19°C showed that fit runners drank an average of 0.7 l in a running time of 150 min, and weaker runners drank an average of 1.7 l in 180 min running time. This coincides with figures in the literature which say that well trained athletes drink an average of 0.4 l/h at 60 kg and poorer runners drink 0.6 l/h at 70 kg body weight (NOAKES, 1993). The fluid intake of over 60 fitness athletes in several marathons in normal outside temperatures (17 to 21°C) fluctuated at three hours running time from 1.2 to 2.3 l.

Drinking in advance before endurance exercise is possible to a **limited degree**. Too much and uncontrolled drinking can lead to premature sweating during exercise. As a matter of principle, however, one should never start when still in a dehydrated state.

During longer lasting endurance exercise, fluid intake should always be coordinated with the necessary carbohydrate consumption.

b) Speed Strength Sports

The speed strength sports include the throwing, putting and jumping events in athletics, sprint events (100 to 400 m running), alpine ski sport, ski jumping, weight lifting among others. Because of the variety of exercise forms it is not possible to make uniform recommendations for drinking behaviour in this sport group. The loss of fluids also varies greatly. In most sports **fluid intake** take place in the breaks or **between starts** (heats, finals). Whether exercise is outdoors or indoors also has an influence. The **quantities drunk** vary individually, but they are **limited** with regard to quantity. Between starts a maximum of 1 l is drunk in breaks. The selection of beverages does not influence performance capacity. Stimulating drinks (cola) with carbohydrates, however, are not disadvantageous. Increasingly preference is given to carbohydrate-electrolyte drinks.

c) Combat Sports

In the classic combat sports of wrestling, judo, boxing and fencing as well as in karate and taekwondo, fluid intake is of great significance especially after competition, or when the weight class is ascertained after weighing (see chapter 3.1.4). When **"making weight"** the athlete tries to lower his training weight to the next lowest weight class, which means reduction of body mass of 4-6 kg is preprogrammed. The athlete structures his **sweating procedures** in such a way that he excretes 3-6 l of fluids in a short period. Through **passive sweating (sauna)** fluid loss only takes place from the **intercellular spaces**. Only by load in greatly insulated clothing does fluid escape from the muscle cells.

The **methods** currently common for **weight reduction** at short notice are restriction of fluid intake, loads with warm clothing and saunas. Rapid fluid loss is dealt with physiologically by the body less well than gradual loss. Whereas in endurance performances a **gradual fluid loss of up to 4%** can be handled without performance restrictions, this is not possible in the case of rapid dehydration. With a rapid fluid loss of 4% in weight class sports there is a high probability that a **reduction of strength performance capacity** will occur. With rapid dehydration athletes in the these sports often achieve the opposite. A low calorie diet of 1,200 to 2,500 kcal/day (reduction diet) with restrictive fluid intake works better. Drastic "weight making" through dehydration only is better avoided, its effect on performance capacity is uncertain and it is damaging to health if repeated frequently.

d) Game Sports

The best known game sports are soccer, tennis, team handball, volleyball, basketball, ice hockey, hockey, water-polo and table tennis. They are generally team sports; in extreme cases only two people play. Because of higher room temperatures, lack of air circulation and sometimes high humidity, **indoor games** lead to rapid and **major heating** up of the players and thus to high water loss via the respiratory tract and sweat. Meanwhile it has become accepted practice that in the short **game breaks** players regularly consume **electrolyte beverages** or carbohydrate drinks (see chapter 3.2.2). A combination of carbohydrates with electrolyte solutions is best when a **game duration** is more than an hour. Outdoor games are less exerting than indoors from a fluid balance point of view, though in summer they are subject to major weather influences. At outside temperatures above 25°C it is necessary to prepare for "heat games" with correspondingly high fluid intake. Here it is necessary to provide fluids for the whole team in advance. Above all, additional water for cooling body parts must be provided.

e) Technical Sports

The technical sports, also called technical-acrobatic sports, include gymnastics, rhythmical sports gymnastics, water-ski jumping, figure skating, ski jumping, ski acrobatics, shooting and sailing among others. When practising these generally short duration sports, **fluid loss** through sweat is **low**. On average sweat loss is 100 to 500 ml. The greatest fluid loss usually takes place in the warm-up phase. Athletes have practically no problems with their fluid balance, and even in a dehydrated state can still be fit for short periods (see chapter 3.2.2).

In gymnastics and rhythmical sports gymnastics, orientation towards as low a body mass as possible has the effect that the girls usually drink too little. Often their coaches instruct them to behave this way. To maintain the **kidney functions** it must be ensured that in spite of the low body mass of these athletes they have a **minimum intake** of 40 to 50 ml of **fluid** per kg of body mass each day. At 35 to 45 kg of body mass that means consumption of 1.4 to 1.8 l of fluid per day. As fluids milk, fruit juices and mineral water should be given preference.

5.4 Fluid Intake During Altitude Training

Ample fluid intake at medium and high altitudes is of great importance for maintaining performance capacity. Three times **more breathing** in the first 2-4 days at a high altitude stay leads to a **rapid fluid loss via the respiratory tract**. The fluid loss is increased by the lower steam pressure in the respiratory tract caused by lower air pressure at increasing altitude. The fluid deficit is first noticed as increasing **dryness in the mouth and the respiratory tract**.

Increased **water loss** via **breathing** and **sweat production** during exercise lead to a decrease in the fluid blood components. The resulting thickening of the blood or haemoconcentration represents an increasing danger to health and performance capacity. When the haematocrit rises above 55% there is danger of embolism.

An important practical **control measure** in altitude training, or in mountain climbing, is to check the fluid and food deficit through **daily weighing of body mass**. Drastic weight loss is always a sign of serious disturbances in the fluid balance (BERGHOLD/PALLASMAN, 1983). At an altitude of 5,000 m about six litres of fluid is lost each day via the respiratory tract alone. This water excretion does not mean a loss of minerals. Minerals in the body are only reduced via sweat and urine, not via the respiratory tract. In addition to **weight monitoring**, at high altitudes the **quantity and colour of the urine** should be constantly checked. Dark or highly concentrated urine is an indication of a fluid deficit. If body mass decreases by 5 kg, in addition to deliberate fluid intake (melted snow) simultaneous consumption of minerals (mineral preparations) is necessary. Melting snow practically means consuming distilled water. If this is not taken into consideration a mineral deficit and a drop in performance are automatic.

5.5 Fluid Intake in the Heat

During sporting loads in heat (about 25°C), fluid requirements are influenced by **load intensity, outside temperature, humidity, wind and sunshine**. In a dehydrated state longer lasting exercise in heat must not be undertaken. To maintain performance capacity the flowing properties of the blood must correspond to normal conditions. The **haematocrit value** should not rise above 50%. Normally in an athlete with 70 kg body mass sweat loss during an endurance run in normal outside temperatures is 1.2 l/h. During exercise in heat sweat loss increases to over 1.5 l/h of exercise and is sure to lead to major dehydration. Dehydration is inevitable in heat loads because fluid intake is always less than the loss via sweat and the respiratory tract.

A **fluid loss** of 1-2% of body weight (0.7-1.4 l of a 70 kg person) does not influence performance capacity. Only gradual dehydration of 4 to 5% definitely has negative effects on sporting performance capacity (WYNDHAM/ STRYDOM, 1986). Sporting load in heat causes water losses both from the cellular and the extracellular spaces in the body. This water loss from two cell spaces is the reason rehydration after sporting loads takes longer than after visiting a sauna. In a normal sauna bath only extracellular water is sweated.

A **control method** for estimating the water deficit is **weighing**. Weight loss of 1 kg after heat loads usually means 85% fluid loss, i.e. 850 g of water and electrolytes and glycogen breakdown of 150 g.

Water loss can be estimated by exact weighing before and after a heat load (BÖHMER, 1981):

$$\text{Loss of body water in \%} = \frac{\text{BM x B\%}}{\text{EM x A\%}}$$

BM = Difference in body mass before and after load
EM = kg body mass
B% = Ratio of water loss to loss of body mass (estimation value 80%)
A% = Ratio of water mass to body mass in % (estimation value 60%)

Example: If a 70 kg athlete with a 60% ratio of water to total body weight loses 2 kg as a result of exercise, then 80% of the weight loss is water. The calculation is:

$$\frac{2 \text{ kg} \times 80\%}{70 \text{ kg} \times 60\%} \times 100\% = 3.8\% \text{ body water loss}$$

If dehydration through heat load of several hours is 5% of body weight, this means 3.5 litres of fluid loss and with certainty a considerable hindrance to performance capacity (see Fig. 7). Compensating a great loss of fluid can take 2-4 days. **Compensation of the fluid deficit** occurs differently in the individual body spaces. The volume of the thigh and the lower leg (muscles) reaches its original state more quickly than the subcutaneous tissue (skin fold thickness). Water only completely returns to the subcutaneous fatty tissue after 3-4 days. Strenuous and dehydrating loads leave signs in an athlete's facial features which can be seen for a long time. During and after heat loads fluid intake is only effective if in the form of an **electrolyte solution**. After greatly dehydrating loads there is **excess fluid consumption**. On the third day the excess fluid intake is corrected hormonally. A drop in the increased aldosterone which resulted from exercise makes the high rate of fluid excretion possible. The athlete notices it in the form of a sudden increase in urine production not in keeping with the amount of fluid drunk. Original body weight returns. **Aldosterone**, which rises by five to twenty times during long lasting exercise, creates a state of hypervolaemia, a dilution of the blood. Through this metabolism functions and the release of oxygen to the tissue is secured. Independently of this regulation a loss of water develops in the intracellular spaces and in the muscles (muscle cells). Heat loads always cause a fluid deficit because **water absorbtion is limited** to a maximum of 1.5 l/h of exercise. Thus water loss via sweat and breathing is usually greater than the possibility to compensate by drinking. An added difficulty is that during exercise athletes cannot drink unlimited quantities. Large quantities of electrolyte solutions can cause **stomach upsets** and/or diarrhoea. At over 800 ml fluid intake per hour runners get an uncomfortable feeling in the stomach; at 1.6 l/h diarrhoea occurs. The **upper limit of digestibility** of fluid quantities in great heat can be assumed to be 1.2 l/h. In cycling consumption of a greater quantity of fluid is less disruptive than in running.

Recommended Vitamin Supply for Untrained Persons

(DGE, Deutsche Gesellschaft für Ernährung/German Association for Nutrition) 1991

Vitamins	Men		Women	
	15-18 Years	19-25 Years	15-18 Years	19-25 Years
Vit. B_1 (mg)	1,6	1,4	1,3	1,2
Vit. B_2 (mg)	1,8	1,7	1,7	1,5
Vit. B_3 (mg) (Niacin)	20	18	16	15
Vit. B_6 (mg)	2,1	1,8	1,8	1,8
Folic Acid (μg)	300	300	150	150
Vit. B_{12} (μg)	3	3	3	3
Vit. C (mg)	75	75	75	75
Vit. A (μg)	1000	1000	900	800
Vit. D (μg)	5	5	5	5
Vit. E (mg)	12	12	12	12

Table 32

6 Vitamin Ingestion in Sport

Vitamins are active substances **essential** for the body, which is constantly dependent on their supply. Vitamins are needed by the body for vital functions. Without vitamins, growth and important life functions are not possible. The body does not produce vitamins itself, or only in small quantities. Vitamins must be regularly supplied with food, either as complete vitamins or as provitamins. **Provitamins** are transformed by the body into vitamins, e.g beta carotene into vitamin A. **Vitamins** are neither building material nor energy suppliers. They function as **coenzymes** or **hormone-like** substances. Only a small quantity of vitamins is required.

The recommended **vitamin requirements** for untrained persons are usually not enough for people practising sports who place themselves under more than 20 hours of load per week. The higher energy utilisation and increased excretion of vitamins via sweat are objective factors for higher requirements of competitive athletes. For competitive athletes additional ingestion of vitamins is now normal. All forms of **vitamin ingestion**, even in high doses, **do not lead to direct improvement in performance**. The adaptive processes and regeneration triggered by the training stimulus only occur without disruption if there is an optimal supply of vitamins. Because there are no reliable figures on exact individual vitamin requirements, in practice a certain surplus is taken. On the basis of current knowledge, the common criteria of deficiency of a vitamin which can lead to symptoms of bad health or illnesses are not a valid argument against an ample supply of vitamins.

Ingestion of vitamins is influenced by numerous factors. The **factors influencing vitamin requirements** include the level of physical and sporting load, the gastro-intestinal functions, various illnesses, consumption of luxury items, pregnancy, growth, the state of regeneration etc. The **vitamin requirement recommendations** of the German Asociation for Nutrition (DGE), or the U.S. National Research Council with its "Recommended Daily Allowances"(RDA), work with **safety margins** and are probably sufficient for 98% of the population (Table 32). In addition, recommendations on vitamin consumption vary considerably from country to country because of the dietary situation and climatic factors. There are **no differences** in **effect** between naturally ingested and industrially manufactured vitamins, except perhaps when a natural product contains additional active substances with a vitamin-like character.

Division of Vitamins Officially recognised by the WHO*

Vitamin Name	Description	Solubility
Vitamin A	Retinol	Fat
Vitamin B_1	Thiamine	Water
Vitamin B_2	Riboflavin	Water
Vitamin B_3	Niacin	Water
Vitamin B_5	Pantothenic Acid	Water
Vitamin B_6	Pyridoxine	Water
Vitamin B_{12}	Cobalamin	Water
Vitamin C	Ascorbic Acid	Water
Vitamin D	Calciferol	Fat
Vitamin E	Alpha Tocopherol	Fat
Vitamin K	Phyllochinon (K1) Menachinon (K2)	Fat
Unofficial Vitamins		
Vitamin F_1	Omega-3 Fatty Acids	Fat/Water
Vitamin F_2	Omega-6 Fatty Acids	Fat/Water
Vitamin H	Biotin	Water
Vitamin B_y	Pteridin	Water
Vitamin J	Choline	Water
Vitamin M	Folic Acid	Water
Vitamin O	Carnitine	Fat/Water
Vitamin P	Bioflavenoids (e.g. Rutin)	Water
Vitamin Q	Ubichinon	Fat

Table 33: WHO = World Health Organisation

It may be of interest that in the division and assignation of active substances to the vitamins there is still no uniformity internationally. To date the World Health Organisation (WHO) only recognises eleven vitamins as **true vitamins** (see Table 33).

Still classified as **unofficial vitamins** are F_1 (omega-3 fatty acids), F_2 (omega-6 fatty acids), H (formerly Bx/biotin), B_y (pteridin), J (choline), M (folic acid), O (formerly B_{11}/carnitine), P (C_2/bioflavenoids, e.g. rutin) and Q (ubichinon).

The classic division of the vitamins has been brought into confusion by new scientific data through the addition of new active substances important to health such as antioxidants, bioflavenoids, essential fatty acids, provitamin A (beta carotene) among others. The bioflavenoids for example have taken on a central role in the prevention of cardiovascular illnesses (HERTOG et al. 1993). The bioflavenoids are probably the central substance in the "French paradox", or the special Mediterranean diet, which has a protective effect on the cardiovascular system because of its strong antioxidative effect.

Metabolic Effect of Vitamins

Vitamins are divided into water soluble and fat soluble vitamins according to their **solubility**. The fat soluble vitamins are A, D, E and K. In addition there are vitamin Q and the vitamins F_1 and F_2 and O which are both fat and water soluble.

Vitamins B_1, B_2, B_6, B_{12}, folic acid, pantothenic acid, niacin, biotin and C are only **water soluble**. Of those vitamins which the nomenclature has not yet clearly identified as such, the vitamins H, B_y, J, M and P are also water soluble.

Vitamin A

Occurrence

The term **vitamin A** is a collective term. **Vitamin A_1** is **retinol** and **vitamin A_2** is **3-dehydroretinol**. The effect of the A vitamins can be clearly distinguished from the provitamin of vitamin A, **beta carotene**. Vitamin A does not occur in plants, but only in **animal** foodstuffs. **Plants** can only produce the **preliminary** form or provitamin, beta carotene. In the intestines this provitamin is enzymatically transformed into vitamin A. Carrots are the main supplier of this provitamin (carotenoids)

Amongst animal products, liver, butter and egg yolk have the greatest vitamin A content. Other vitamin A sources are cheese, margarine and sea fish.

Function

The main **effects** are participation in growth and the differentiation of skin and mucous membranes as well as vision (Table 34). The light-sensitive pigment (rhodopsin) in the eye is produced from vitamin A. Probably retinol regulates gene expression of the growth hormone. Reproduction is dependent on the supply of vitamin A. Vitamin A is needed for pregnancy and the development of the male sexual hormone testosterone. The body has certain **vitamin A stores**. The main stock of vitamin A, 95-99%, is in the liver. In case of deficiency the store is only emptied after several months.

Requirements

Requirements are given in **retinol equivalents** and are complemented with beta carotenoids. The previously common International Unit (I.U.) is the equivalent of 0.3 μg of retinol. The **daily requirement** is accepted to be 1 mg/d. It is recommended to cover vitamin A requirements with 1/3 retinol and 2/3 beta carotene. This mixture is sensible because beta carotene is an important **oxygen free radical catcher** and has a strong **antioxidative effect**.

In **competitive sport** a higher **vitamin A requirement** is necessary; it exceeds the requirements of untrained persons by a factor of 4 to 5, i.e. 4 to 5 mg/d of retinol are needed, complemented additionally with beta carotene. The normal vitamin A concentration in the blood is >20 μg/dl vitamin A and >40 μg/dl carotene. If the blood concentration goes below 10 μg/dl, then the vitamin A stores are almost empty.

One of the first signs of a vitamin A deficiency are problems with twilight vision. Later drying out of the hair and mucous membranes follows. To treat **deficiency symptoms** or illnesses the oral dosage is increased to 2,000 to 4,000 μg. Because of its fat solubility vitamin A can be overdosed. (Table 35). An overdose can occur when supply is ten times the requirement. Beta carotene cannot be overdosed; as a result of oversupply (carrots) a yellowing of the skin and mucous membranes occurs.

Vitamin D

Occurrence

Vitamin D consists of several active substances which together are called **calciferols**. With vitamin D also there are **provitamins**, ergosterol and **7-dihydrocholesterol**. 7-dihydrocholesterol is transformed in human skin and through **UV radiation** (sun, mountain sun) into the effective **vitamin D_3**.

Physiological Effects and Requirements of Fat Soluble Vitamins in Competitive Sport

Vitamins	Effect	Intake per Day
Vitamin A (retinol)	Protective substance: Skin, Mucous membranes Vision (rhodopsin development) Protein synthesis	2-5 mg Retinol Equivalent/RE (5,000-15,000 IU Vit. A) Maximum dosage: during pregnancy 15,000 IU Vit. A (otherwise 25,000 IU)
Beta carotene (Provitamin A)	Antioxidant; Only about 15% of vitamin A effect	2-4 mg RE; overdose not possible (skin yellowing)
Vitamin E (alpha tocopherol)	Strong antioxidant: Protects unsaturated fatty acids, Vit. A, hormones and enzymes against oxidation. Arteriosclerosis protection	20 - 300 mg
Vitamin D (calciferol)	Growth and bone development	5 - 10 μg Vit. D_2 resp. D_3 Overdose possible above 25 μg (1,000 IU)
Vitamin K (phyllochinon, menachinon)	Blood coagulation, bone metabolism	70 - 140 μg Vit. K1 (phyllochinon)
Vitamin Q (ubichinon)	Electron carrier in the breathing chain, key function for cellular energy development, antioxidative effect with Vit. E, C and P	10-30 mg coenzyme Q_{10}

Table 34

Vitamin D_3 is mainly found in **animal** products (butter, cheese, liver) and is especially plentiful in sea fish (herring, salmon) and fish liver oils (cod liver oil). Because normal mixed diets are only a small source of vitamin D_3, numerous **foods** (margarine, butter) are **enriched** with it. Thus prevention of rachitis (rickets) in childhood is achieved. For adults the body's own development of vitamin D from 7-dihydrocholesterol in cholesterol is the main source. The transformation of the D vitamins occurs in complicated metabolic processes. For adults the vitamin D_3 development (cholecalciferol) stimulated by UV sun radiation is sufficient. The vitamin D_3 of the skin is transformed in the kidney to calcitriol (1.25-dihydroxycalciferol) via 25-hydrocholecalciferol developed in the liver. Calcitriol production is finely regulated and is adapted to the body's calcium requirements.

Functions

Vitamin D is the preliminary stage for **hormone-like active substances** which are involved in the calcium and phosphate balance. The individual D hormones work in the intestines, the kidneys and the bones. The D vitamins are decisive for the **mineralisation** of the bones. When there is a vitamin D deficiency the concentrations of calcium and phosphate in the blood drop. The consequences of insufficient vitamin D are rachitis in children and various forms of softening of the bones (osteomalacia) in adults. Lack of sunlight due to masking clothing in childhood and adulthood is still the main cause of bone development problems in certain countries.

Requirements

The requirements are given in I.U. or μg 1 I.U. equals 0.025 μg of vitamin D_2 or D_3 (or 1 mg of $D_{2,3}$ equals 40,000 I.U.).

In adulthood the vitamin D requirements are mainly secured through **self-synthesis** (7-dehydrocholesterol of the skin) when there is **sun exposure**. Thus the D vitamins have a special position amongst the vitamins. The DGE recommends a supply of 5 μg/day in adulthood. Athletes who train outdoors or in warm climate zones have no problems with supply.

Through the exposure of 1 cm^2 of skin 10 I.U. (0.25 μg) of vitamin D an hour are developed from 7-dihydrocholesterol (FRIEDRICH, 1987). For athletes in **indoor sports**, swimmers or those training with protective clothing, intake of 10 μg/day is sufficient, especially in the more or less sunless winter months. For **rachitis prevention** 12.5 to 25 μg (500-1,000 I.U.)/day of vitamin D_3 are necessary. Therapeutically the dosage is much higher.

Excessive vitamin D consumption of 1,000-3,000 I.U. over several months has a toxic effect, leading to development of stones and calcification of soft parts. Because of the hormone-like effect of the D vitamins there was a dispute in the Twenties as to whether they are vitamins or hormones.

Vitamins	Daily Requirements: Untrained Persons*)		Daily Requirements: Athletes		Minimum Toxic Dose	
A (retinol)	5000 IU	(1.5 mg)	13000 IU	(4.47 mg)	25-50 000 IE	(7.5 - 15 mg)
beta carotene	3 mg		4.5 mg		30 mg	
D_3 (calciferol)	400 IU	(10 µg)	800 IU	(20 µg)	5000 IU	(1.2 g)
E (tocopherol)	10 mg		50 mg		1.2 g	
C (ascorbic acid)	60 mg		300 - 500 mg	2 g[2]	5 g	
B_1 (thiamine)	1.5 mg		7 mg	8 mg[1]	300 mg	
B_2 (riboflavin)	1.8 mg		8 mg	12 mg[1]	1 g	
B_3 (niacin)	20 mg		30 mg	40 mg[1]	1 g	
B_4 (folic acid)	300 µg		400 µg		400 mg	
B_5 (pantothenic acid)	(10) mg	not fixed	20 mg		10 g	
B_6 (pyridoxine)	2.1 mg		10 mg	15 mg[1]	2 g	
B_{12} (cobalamin)	3 µg		6 µg		20 mg	
K (phyllochinon)	80 µg		150 µg		2 g	
H (biotin)	0.1 mg		0.3 mg		50 mg	
Q (ubichinon)		not fixed				

*Table 35: *) Recommendations of the NRC/RDA (National Research Council, Recommended Daily Allowance 1989 (USA) and DGE (German Association for Nutrition) 1991*

1 Strength - Speed strength athletes 2 Altitude training

Vitamin E (tocopherol)	
Effect:	***Very effective lipophile antioxidant.*** Together with Vit. Q it makes up the first antioxidative line of defence in the lipophile layer of the cell membrane. It transfers electrons (free radicals) to Vit. C (second antoxidative line of defence). In the vitamin Q-E-C cycle, Vit. C. takes the free radicals from Vit. E. Vit. E protects unsaturated fatty acids, Vit. A, hormones and enzymes against oxidation.
Daily Requirements:	3 - 15 mg/day Normal ingestion 12 - 20 mg Untrained persons 20 - 40 mg Children, fitness athletes 100 - 200 mg Competitve athletes 300-1,000 mg Elite athletes with great muscle use
Foods with high Vit. E content:	Wheat germ, vegetable oils, margarine, grain, natural rice, oatflakes, vegetables (asparagus, spinach, Brussel sprouts, broccoli), potatoes, eggs, milk
Medicines:	5 mg to 500 mg alpha tocopherol

Table 36

Vitamin E

Occurrence

Vitamin E or tocopherol is a collective term for various substances which have similar effects. The main effect comes from **alpha tocopherol**. Tocopherols are produced in plants. **Vegetable oils** are especially rich in tocopherols. Wheat germ and sunflower oil have a high vitamin E content (215 resp. 56 mg/100 g). After these oils hazelnuts and walnuts contain the greatest quantities of vitamin E (20 resp. 26 mg/100 g). Meat, fish and dairy products, however, also contain vitamin E. As a result of **industrial processing** of plants (refining) about 1/3 of the natural vitamin E content is lost. Slow growing and green plants have a higher vitamin E content than fast growing green plants other than green. Grain and grain products are the **main reserve** for vitamin E ingestion.

Functions

The effect of the **four tocopherols** (alpha, beta, gamma and delta tocopherol), and the four **tocotrienols** is diverse. Their common mechanism is the strong **antioxidative effect** on redox (oxidation reduction) systems (Table 36). Vitamin E protects the unsaturated fatty acids (linoleic acid and linolenic acid), vitamin A and various hormones and enzymes against oxidation. Furthermore, vitamin E influences **protein synthesis, the immune function and the neuromuscular system.**

For performance capacity in sport the involvement in **electron transport** to and in the breathing chain, and thus in **energy production**, is of significance. Thus vitamin E is necessary for the effective operation of aerobic energy production. In addition to involvement in **aerobic metabolism**, fat soluble alpha tocopherol protects the membrane structures, which consist of unsaturated fatty acids, against the influence of metabolism radicals. Vitamin E protects the **cell membranes** against destruction by oxygen radicals and is thus necessary for the **stabilisation** of the structure of these. The natural antioxidative protective system of the body can be overtaxed by great unaccustomed loads. A vitamin E deficit reduces the activity of the antioxidative enzymes in loaded muscle (BERG et al. 1987). Ample vitamin E content increases **muscular load tolerance** and prevents **sore muscles**. The absorption of vitamin E in the intestines is dependent on the fat content of food, i.e. it increases with more fat. Several grams of vitamin E can be stored in the fatty and liver tissues.

Requirements

The exact requirements are not known so figures on desirable **ingestion** vary greatly. The serum concentration is about 1 mg/100 ml. For adults daily ingestion of 15 to 20 mg (or 15-20 I.U.) of alpha tocopherol is recommended. Requirements rise with increased ingestion of polyunsaturated fatty acids. For the ingestion of 0.6 g of unsaturated fatty acids, 1 mg of vitamin E is needed. A low vitamin E level seems to be a **risk factor** for cancer and heart attacks. The dosage of the various tocopherols is related to the **biological effect** of alpha tocopherol. 1 mg of alpha tocopherol equals 1.0 Alpha Tocopherol Equivalent or 1.49 I.U.. In sport, and in particular in competitive sport, the **average recommendations** are not sufficient (see Table 36). As a rule they are a factor of 10 or 20 higher than recommended for untrained persons. Doses of 400 I.U. (596 mg) can be handled without side-effects. Ingestion of 3 g/day over ten years without problems has been described. Because of their higher energy utilisation, greater muscle use and raised aerobic metabolism, sport practitioners always have greater vitamin E requirements than untrained persons.

Vitamin B_1

Occurrence

Vitamin B_1 (thiamine or aneurin) is found in both **animal** and **vegetable** foods. **Grain products** (wheat, rye, oatflakes), corn and rice have a high vitamin B_1 content (0.4 to 0.6 mg/100 g). The content in **pork**, however, is even greater, at 0.9 mg/100 g. Beef has only 1/3 of the B_1 content of pork. Various **vegetables** (e.g. peas with 0.3 mg/100 g, potatoes and carrots with 0.1 mg/100 g each) are carriers of thiamine in food. When grain is ground or rice polished more than half the thiamine is lost because it is mainly contained in the outer layers.

Vitamin B_1 avitaminosis is known as **"beriberi"**. In its "wet" form the symptom is the development of oedema. The "dry" form has symptoms of muscle paralysis, psychological disturbances and memory problems. In industrialised countries "beriberi" is very rare. Beriberi means sheep's gait, derived from the stiff gait of sheep and indicating nerve disturbances. High alcohol consumption leads to a thiamine deficiency and in an advanced stage alcoholics have problems walking. Diabetics can also have a thiamine deficit, usually in connection with a magnesium deficiency.

Functions

Vitamin B_1 is a component of **enzymes** in anaerobic and aerobic **carbohydrate metabolism**. As a water soluble vitamin it is unstable in heat and is destroyed by cooking. The breakdown of pyruvate to activated acetic acid (acetyl-CoA) is dependent on vitamin B_1. With increasing energy utilisation the B_1 requirement rises, per 1,000 kcal/day 0.5 mg of thiamine are needed. B_1 is also necessary for the breakdown of alpha ketoglutarate in the citric acid cycle. The effect in the nervous system is not yet known exactly, but thiamine is necessary for the development of acetyl choline, a nerve transfer substance. The breakdown of the branch chain amino acids valine, leucine and isoleucine in gluconeogenesis is dependent on vitamin B_1.

Requirements

The normal thiamine requirements are given as **0.5 mg/1,000 kcal/day** or 1.2-1.4 mg/day. A study with competitive athletes showed that their supply is insufficient, for 16% were under the requirements for untrained persons and 50% under those of a control person (ROKITZKI et al. 1994). Only marathon runners had a satisfactory blood level in this study because they deliberately took thiamine. The **plasma concentration** of free thiamine is 1 μg/100 ml (1 I.U. of thiamine equals 0.003 mg). An increase in the number of calories in food does not automatically lead to higher ingestion of this vitamin. Many food concentrates on a carbohydrate basis are low in vitamin B_1.

Physiological Effects and Dosages of Vitamins in Competitive Sport

Vitamins (water soluble)	**Effect**	**Dose per Day**
B_1 (thiamine)	Aerobic energy metabolism, heart and nerve functions	6-10 mg
B_2 (riboflavin)	Anaerobic and aerobic energy metabolism	6-12 mg
B_6 (pyridoxine)	Protein metabolism, Antioxidant	6-15 mg
Niacin	Energy metabolism, biosyntheses	20 - 40 mg
Pantothenic acid	Aerobic energy metabolism Antioxidant	4 - 7 mg
Biotin	Fatty acid synthesis, gluconeogenesis, immunity through T and B cells	50 - 100 μg
Folic acid	Cell development, DNA synthesis, immune system, blood coagulation	200 - 400 μg
B_{12} (cobalamin)	Cell development, DNA synthesis, L-carnitine synthesis (fatty acid oxidation), immune system	2 - 6 μg
C (ascorbic acid)	Antioxidant, infect defence in immune system, L-carnitine synthesis	300 - 500 mg

Table 37

The absorption limit in the intestine for water soluble thiamine is about 15 mg/day. In **competitive sport supplementation with B_1 is advisable**, because in addition to increased consumption in metabolism, this vitamin is also excreted in sweat and urine. For competitive athletes training more than 20 h/week, 6 to 10 mg/day are recommended (Table 37). In high load phases higher doses are possible.

Through the development of a fat soluble vitamin B_1 derivate, **benfotiamin**, considerably greater ingestion in the intestines can be achieved. The ingestion of 1 g/day of a fat soluble thiamine derivate for four days did not influence sporting performance capacity (WEBSTER et al.1997).

Whereas in case of deficiency with water soluble thiamine 20-30 mg can be taken daily; with the fat soluble version 100 mg are possible.

Vitamin B_2

Occurrence

Vitamin B_2 or **riboflavin (lactoflavin)**, is widespread in the animal and plant kingdoms and in solution is yellow-green. The greatest vitamin B_2 content is found in yeast, but this is not of significance from a nutritional physiology point of view. Sufficient intake is achieved from **milk** and **dairy products** (0.2-0.3 mg/100 g). In **meat** vitamin B_2 is plentiful (0.2 mg/100 g). Occurrence in cattle liver is considerably higher at 3 mg/100 g. It is found in quantities of 0.1 to 0.2 mg/100 g in peas, beans and cabbage. The grains of wheat, corn and rice contain 0.1 mg/100 g of vitamin B_2. Surplus vitamin B_2 is **not stored**. The small reserves last 2-6 weeks. Riboflavin is alkaline and light sensitive, and up to half of it is lost through incorrect storage and food preparation.

Functions

Riboflavin is the **coenzyme** of a large number of reducing substances, which are called flavoproteins or **flavoenzymes** owing to their yellow colour. Riboflavin is involved in the breathing chain and is necessary for hydrogen transfer. As a component of the **enzyme system** of the **breathing chain** in the **mitochondria** it is always needed for the **aerobic energy metabolism**. A deficiency of Riboflavin can lead to a secondary deficit of vitamin B_6, pantothenic acid and possibly also folic acid and niacin (BÄSSLER, 1992). The interaction of the B vitamins can be seen in Fig. 25.

Requirements

For untrained persons the requirements are 1.8 to 2.5 g/day. The plasma concentration of riboflavin is 2-4 μg/100 ml. Sport training increases riboflavin requirements. With increasing energy consumption the requirements rise. **Ingestion** should be at least 0.6 mg/1,000 kcal/day. For great training loads 6 to 12 mg/day are recommended (see Table 37). In competitive sport there is no evidence of a vitamin B_2 deficit when a balanced diet is adhered to (ROKITZKI et al., 1994). The human intestinal flora can develop riboflavin. Riboflavin is not toxic even in high doses.

A training run of top class triathlete Lothar Leder in the company of Federal coach S. Grosse

The men's start in duathlon

Nourishment during the run

European triathlon champion Dr Rainer Müller quenching his thirst

Duathlon world champion Norman Stadler with plans for the future

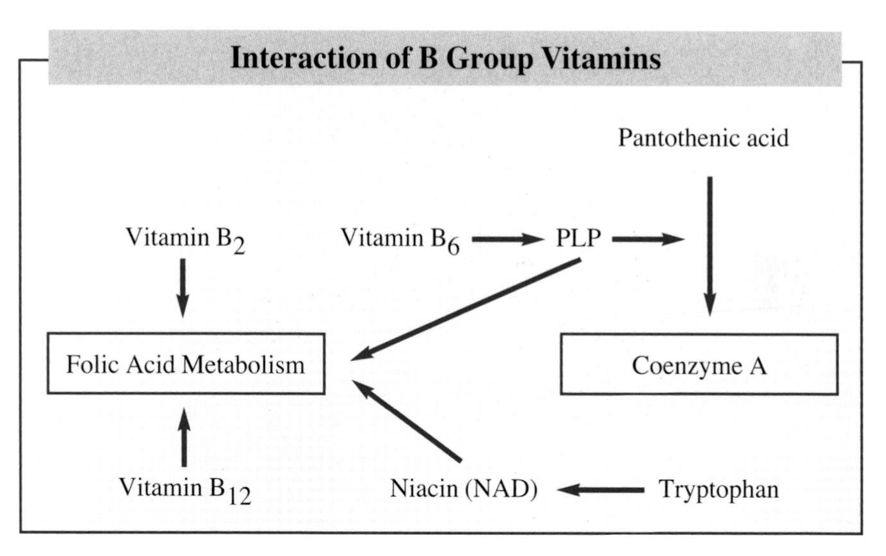

PLP = Pyridoxal phosphate
NAD = Nicotinamide-adenine-dinucleotide

Modified from Bässler (1992)

Fig. 25: Influence of B group vitamins and the amino acid tryptophan on folic acid metabolism and their mutual regulatory dependence in influencing aerobic metabolism (Coenzyme A). Modified from Bässler, 1992.

Vitamin B_6

Occurrence

Vitamin B_6, or **pyridoxine** (adermine), is widespread in nature. In **fish** and **meat** it is found in quantities of 0.4 to 0.8 mg/100 g. The next most importance occurrence is in **grain**, corn and rice (0.2 to 0.6 mg/100 g). Smaller quantities are found in fruit and vegetables (0.1 to 0.3 mg/100 g). **Loss during food preparation** is 20 to 40% so that underheated grain products, grain germs and vegetarian whole foods contribute to a sufficient supply.

Function

Vitamin B_6 is the **coenzyme** of many enzymes in **amino acid metabolism**. Vitamin B_6 is made up of three pyridoxines (pyridoxal, pyridoxamine and pyridoxol). These are the coenzymes of numerous enzymes. **Protein synthesis** (organ growth, muscle development, muscle regeneration) is tied to the presence of vitamin B_6. Pyridoxin is known as an effective antioxidant (see Table 37).

Requirements

Vitamin B_6 requirements depend on **protein utilisation** and increase with increasing protein ingestion as well as fatty acid ingestion. The increased requirements are caused by the involvement of this vitamin in metabolism of the **amino acids**. For untrained persons ingestion of 1.5 to 2.5 mg/day is considered desirable. Recent findings in competitive sport show that over 30% of athletes **do not get the minimum** amount of 1.6 to 1.8 mg/day (ROKITZKI et al. 1994). After major loads, such as a marathon, there is a loss of over 1 mg of vitamin B_6 (ROTKITZKI et al. 1994). Thus competitive athletes should pay special attention to their supply of vitamin B_6 which, depending on the sport, is about 6 to 15 mg/day. The greatest need for vitamin B_6 ingestion is among **strength athletes**, or where there is increased protein utilisation. **Vitamin B6** is thus seen as the **key vitamin in competitive sport training** and must be available in sufficient quantities for protein metabolism in the loaded muscles. Alcoholism leads to pyridoxine deficiencies. It should be noted that taking the "contraceptive pill" leads to a deficiency so that under this aspect female endurance athletes should take special care to take ample pyridoxine. First indications of deficiency are dry skin, tears at the corners of the mouth and an inflamed tongue. The upper dosage limit should not be above 1,000 mg/day.

Vitamin B_{12}

Occurrence

Vitamin B_{12} or **cobalamin** is only found in animal products. Cattle and pig **liver** contain the most vitamin B_{12} (70 and 30 μg/100 g respectively). The kidneys are also rich in this vitamin. The meat of the muscles contains considerably less (2-3 μg/100 g). Amongst fish, herring has the highest concentration of vitamin B_{12} (10 μg/100 g). Other sources are eggs, cheese and full cream milk (2-4 μg/100 g). **Vegetable foods are free of vitamin B_{12}!**

Function

Ingestion of vitamin B_{12} in the intestines is tied to a physiological component of the gastric juices, the **intrinsic factor**. If this is missing, a deficit arises. Vitamin B_{12} is the only biological substance containing cobalt. In metabolism it is involved in the **reducing systems** of the mitochondria in the breakdown of fatty acids and amino acids. Vitamin B_{12} supports the breakdown of branch chain amino acids and makes it possible to sweep them into the citric acid cycle. Vitamin B_{12} is necessary for **cell development** and **deoxyribonucleic acid (DNA) synthesis**. The body's own prodution of L-carnitine requires vitamin B_{12}.

L-carnitine is necessary to sweep long chain fatty acids through the inner mitochondria membrane. Without this transport the fatty acids could not be broken down. Because vitamin B_{12} is also required for the maturing of blood in the bone marrow, a deficit leads to errors in the production of the erythrocytes which in the end results in oxygen supply problems **(megaloblastenanaemia)**.

Requirements
The requirements of vitamin B_{12} are given as 2 µg/day, for sport practitioners they are higher by a factor of 3 (see Table 37). Therapeutic dosages are 1 mg/day and cover supply for a month. High dosages of vitamin B_{12} only work with certainty as an injection. Through ingestion with food only about 3% is absorbed. **Additional ingestion of vitamin B_{12} is only necessary for vegetarians**. The main source is meat. When large quantities of beer are drunk, yeasts are ingested which can cover about 10% of requirements. The vitamin B_{12} stores hold about 4 mg and last about 3-4 years. Thus deficiencies only develop very slowly.

Biotin (Vitamin H)
Occurrence
The previous names were vitamin B_7 or vitamin H. Biotin is widespread in nature. **The main sources** in food are liver, kidneys, milk and eggs. Whereas in liver 30 to 100 µg/100 g are found, in soy beans there are 60, in chicken eggs 25, in bananas and wheat grain 5-6 and in meat 2-5 µg/100 g of biotin. In plants biotin occurs in water soluble form and in animal organs and yeast in the insoluble (protein bound) water form.

Function
Biotin is a **coenzyme** of several **metabolic reactions** and is necessary for the **key enzymes** of **gluconeogenesis** and **fatty acid synthesis**. It is thus a link between carbohydrate and fat metabolism. The breakdown of branch chain amino acids also requires biotin. Biotin is involved in cellular immune functions (B and T cells).

Requirements
The requirements are about 50 to 100 µg/day and are easily **covered** by a normal balanced diet (see Table 37). Indications of a deficiency are skin problems (dermatitis, shedding hair), muscle pains and drowsiness. When there is suspicion of a biotin deficiency, daily ingestion of 200-1,000 mg of biotin can remove this. Toxic effects are unknown.

Folic Acid (Vitamin M)

Occurrence

Folic acid is found in vegetable and animal foods as a water soluble substance and belongs to the vitamin B complex. **Greens**, tomatoes, grain and liver are **rich** in folic acid. Meat, fish and fruit, however, contain little folic acid.

The folic acid content in the liver is 330 μg/100 g, in chicken eggs it is 78, in salad, beans, spinach, asparagus and tomatoes 73-106, in milk and cheese 20 and in grain 70 μg/100 g.

Function

Folic Acid is an important coenzyme for metabolism of **amino and nucleic acids**. Here they work as accepter and carrier of activated formaldehyde and formic acid. The **production of new cells** requires folic acid. Folic acid is also involved in the **immune function** and **blood coagulation**. Absorption depends on the dosage and increases linearly to quantities of 5,000 μg. The serum concentration of folic acid is representative for the folic acid status in the body.

Requirements

The daily requirements of folic acid are about 3 μg/kg of body mass, so that about 200 μg for men and 180 μg for women should be ingested (see Table 37). The amount of folic acid in normal food, which is consumed as free and bound folic acid (total folate), must be greater than free folic acid; for adults it should be 400 mg/day. In sport a folic acid deficiency is possible, but the effects are not clear. During pregnancy there are especially high folic acid requirements, which the DGE specifies as 600 μg total folate, or 150 mg free folic acid. A folic acid deficiency during **pregnancy** leads to deformities or premature birth. A deficiency is often coupled with a deficiency of vitamin B_{12}, which results in **anaemia**. Drinking wine and spirits contributes to a **folic acid deficiency**, as does a predominantly **fast food diet**. For sport practitioners regular consumption of greens (salad) is recommended. Because of the folic acid stores a deficiency does not have an effect until after four months.

Nicotinamide (Vitamin PP, niacinamide)

Occurrence

Nicotinamide is the amide of nicotinic acid; together nicotinic acid and nicotinamide are called niacin. The former term "PP" factor comes from the skin disease **pellagra** (brown skin) caused by nicotinamide. Pellagra occurs in places exposed to the sun and is accompanied by fatigue, low performance, and inflammation of mucous membranes. These symptoms were in 1735 known in people who only ate corn. The amino acid **tryptophan** is an important **preliminary**

stage of the production of **nicotinic acid**, so that exact nicotinamide requirements are not known. Nicotinamide is contained in all foods except fats and oils. Yeast, liver and muscle meat are rich in nicotinamide. The quantities are 15 mg/100 g in liver and about 5 mg/100 g in meat. Amongst vegetable foods, grain and grain products have the highest concentration (1-5 mg/100 g). In **roasted coffee** the nicotine content is a high 13 mg/100 g so that five cups of coffee already cover the requirements for a day. Niacin, which occurs in corn grain (1.5 mg/100 g) can only be processed by humans after special treatment (boiling in hard water).

Function

Niacin is the **coenzyme** of important substances in energy metabolism (nicotinic acid amide adenine dinucleotide = NAD and nicotinic acid amide adenine dinucleotide-phosphate = NADP). It is thus necessary for the function of the **breathing chain** in the mitochondria, for the running of **glycolysis**, for **fatty acid synthesis** and other important hydrogen transferring metabolic processes. As nicotinic acid can be produced from the amino acid tryptophan in the body, 60 mg of L-tryptophan are equivalent to 1 mg of nicotinamide. The ingestion of large doses of niacin increases carbohydrate oxidation and suppresses fat metabolism (BEEK, 1991).

Requirements

The requirements cannot be ascertained for certain as the body has the option of producing nicotinic acid amide from **tryptophan**. The nicotinamide requirements are estimated at 18 mg/day. The blood plasma level of nicotinamide is 75 mg/100 g.

The production of nicotinic acid amide in the body requires a sufficient supply of folic acid, vitamin B_2 and vitamin B_6. The desirable amount to take is 20 mg of niacin. Normally 0.5-1.0 g of tryptophan and 8-17 mg of nicotinamide are ingested daily with food. So far no data on states of deficiency or higher requirements in sport is known. Therapeutic doses of nicotinamide in states of deficiency are 50-100 mg/day. Doses of a maximum of 1 g/day should not be exceeded because of major vascular dilation caused by nicotinic acid.

Pantothenic Acid (Vitamin B_3)

Occurrence

Pantothenic acid was originally referred to as a **"growth factor"** or "antidermatitis factor". Because these factors occurred **everywhere**, they were summarised under the term pantothenic acid (pantothen = everywhere). Pantothenic acid is found in almost all foods. Liver (7 mg/100 g), offal (2.7 mg/100 g) and meat (0.6 mg/100 g) are rich in it. But quantities of 1-1.6 mg/100 g are also found in wheat grain, eggs, broccoli or cauliflower.

Function

Pantothenic acid is a component of important substrates in **energy metabolism**, such as activated acetic acid (acetyl coenzyme A) and 4-phosphopantethein. As a coenzyme component pantothenic acid is involved in carbohydrate, protein and fat metabolism (pyruvate breakdown, breakdown of branch chain amino acids, processing of ketone bodies among others). Pantothenic acid is an antidote for the Indian arrow poison curare.

Requirements

Assessments of the pantothenic acid requirements are difficult, so that a daily requirement of 8 mg is assumed. With a supply of only 1 mg/day no deficiency symptoms have been observed. Adjusting diet to concentrates leads to lower ingestion of pantothenic acid in industrial nations (under 2 mg/1,000 kcal). **Stress situations** and greater **energy ingestion**, as are usual in sport, increase requirements of pantothenic acid (see Table 37). Where a deficiency is suspected, 10 mg/day can be taken.

Vitamin C

Occurrence

Vitamin C, or **ascorbic acid**, has a long history based on the fact that the effects of vitamins on certain diseases were unknown. The lack of a supply of citrus fruit or vegetables caused numerous participants in expeditions to **contract scurvy**, and right into the 18th century it was the most frequent cause of death among seafarers.

The vitamin C content is high in **citrus fruits** (50-80 mg/100 g), cabbage-related vegetables (45-110 mg/100 g) and liver (25 mg/100 g). Meat contains no vitamin C. The greatest bearers of vitamin C are rose hips (1250 mg/100 g), sea buckthorn berries (450 mg/100 g), blackcurrants (180 mg/100 g), parsley leaves (160 mg/100 g), and paprika (139 mg/100 g). With 17 mg/100 g the **potato** is also an important supplier of vitamin C through its inclusion in the daily diet. In Germany vitamin C ingestion occurs through fresh vegetables, potatoes, fruit and citrus fruits. Increasingly enriched fruit and orange juices, as well as multivitamin drinks, are contributing to meeting requirements. Ascorbic acid is very sensitive to oxidation, so that great losses occur through cooking and storage.

Function

Viatmin C intercepts cell damaging free radicals in the cell membranes and is thus an effective **antioxidant**. In this role it protects vitamins E, A, thiamine, riboflavin, among others, from destruction. Vitamin C is involved in important metabolic processes (microsomal hydroxylisation reactions and oxygenase reactions). Additionally vitamin C influences **iron metabolism** by supporting absorption in the intestines and increasing the stability of **ferritin** (intracelluler storage iron). It can probably influence damage to liver cells or certain forms of cancer through the suppression of mutational reactions (blocking occurrence of nitrosamine) (TANNENBAUM, 1989). Positive effects of vitamin C on the immune system and the hormonal system in the brain (neuroendocrinium) are probable (DEGKWITZ, 1985).

Requirements

Figures on requirements vary. They range from small doses (10 mg/day) for the prevention of scurvy to 2-4 g/day (see Table 35). The DGE recommendation for ordinary citizens has meanwhile been raised slightly (75 mg/day). It is certain that these low dosages are too low for competitive athletes.

In **competitive training** they need a daily dosage of **300-500 mg** of vitamin C. In **stress situations**, which involve an increase in adrenaline and noradrenaline, and during extreme sporting load (e.g. altitude training, heat acclimatisation), the **ingestion** of vitamin C can be even **higher** and reach 1 to 2 g/day. Because vitamin C is water soluble, if too much is taken it can be excreted via the kidneys without problems. There is no agreement on whether a consequential amount is lost via sweat. Overdoses are practically without significance (see Table 35). At amounts above 2 g/day gastro-intestinal problems can occur and constant high ingestion can lead via oxalic acid (breakdown by-product) to kidney stones (oxalate stones). Internationally, recommendations are tending towards higher dosages of vitamin C. The main indications for high vitamin C ingestion in competitive sport are the **antioxidative effect** (cell protection) and the **stabilising effect on the immune system**. In addition to **prevention of infections**, the **reduction of the risk** of heart diseases and not least the **protective effect against cancer**, regular ingestion of large amounts of vitamin C is also recommended for untrained persons (BÄSSLER et al. 1992).

For the **prevention of infections in sport** intravenous shots of 1-2 g of vitamin C/day can be useful. **High doses of vitamin C do not increase sporting performance capacity**.

Symptoms of vitamin C **deficiency** are **decreasing physical performance capacity**, increasing fatigue, increased need for sleep, joint pains, spontaneous capillary bleeding and inflammation of mucous membranes. There can also be a deficiency if the body has adjusted itself to a certain high dosage and the supply necessary to maintain this higher balance stops. Temporarily symptoms of a relative deficiency can occur in this case.

7 Mineral Ingestion in Sport

Minerals are **inorganic substances** which are indispensible for maintaining life. They are the supporting and hardening substances for the development of the skeleton and the teeth. Securing the **mineral balance** is necessary for the functions of the muscles and other organs (Table 38). Mineral deficiency can lead to serious functional disruptions, especially when the body is under load. Numerous physiological functions, such as osmotic pressure, nerve impulse transmission, triggering of muscle contraction, maintaining enzyme activities among others are not possible without certain quantities of minerals.

For sport it is not so much the absolute mineral balances which are important, but rather those available at the time of loading. The minerals which mainly carry out their functions as electrically charged particles and dissociate into anions and cations are the electrolytes. The most important **electrolytes** in the blood plasma are: Na^+, K^+, $Ca2^+$ and $Mg2^+$. Because they become cathodes they are also called **cations**.

Together with the **anions** (bicarbonate, chloride, phosphate, sulphate, organic acids and proteinate) the **cations** ensure ion equilibrium in the body. The ions (cations and anions) can penetrate the cell membrane walls and the vessel walls. Ion transfer is either passive in the direction of lower concentration or active against the concentration, drop through the ion channels of the cell walls via specific pumping systems, such as e.g. the **potassium-sodium pump**.

Increased calorie content in food does not automatically increase the quantity of minerals. Minerals are also called **micronutrients**, analogous to vitamins. Thus the micronutrient content in food ingested is of nutritional physiological significance. With increasing quantities of micronutrients the quality of food increases. The **quality of nutrient density** is based on the quantity of micronutrients consumed per 1,000 kcal. **Nutrient density is the quotient of nutrient content per 1,000 kcal** and the **recommended ingestion** of micronutrients per 1,000 kcal. In this respect a high calorie fast food diet would have a lower nutrient density than a whole food diet.

Because there continues to be uncertainty about real vitamin and mineral requirements, over half of all athletes, especially in the USA, take food supplements with minerals and vitamins in dosages that are not specifically targeted (BURKE et al., 1993).

The necessity of supplementation is real when regular weekly training load exceeds 15 hours.

Minerals in Sport

Mineral	Daily Requirements		Minimum Toxic Dose
	Untrained Persons*)	Athletes	
Salt	8 g	15 g	> 100 g
Potassium	2,5 g	5 g	12 g
Calcium	1,0 g	2 g	12 g
Phosphorus	1,2 g	2,5 g	12 g
Magnesium	0,4 g	0,6 g	6 g
Iron	18 mg	40 mg	> 100 mg
Zinc	15 mg	25 mg	500 mg
Copper	2 mg	4 mg	100 mg
Fluoride	2 mg	4 mg	20 mg
Iodine	0,15 mg	0,25 mg	2 g
Selenium	70 µg	100 µg	1 mg
Chromium	100 µg	200 µg	2 mg

*Table 38: * RDA (1989) and DGE (1991) recommendations*

Sodium

Sodium is mainly stored in the **body fluids**. Concentration within the cells is considerably lower. The physiological functions of sodium are diverse. It plays a key role in maintaining the **body's water balance**. The sodium concentration influences **blood pressure, osmotic equilibrium**, the **acid-alkali balance** and **muscular sensitivity**. The body stores about 80 g of sodium. Sodium is constantly lost through **sweat**. In industrialised countries salt (NaCl) ingestion is plentiful because of the way food is prepared.

Mineral Deficiency in Competitive Training

Deficiency	Indications	Recommended Foods
Magnesium Serum concentration: < 0.75 mmol/l	Calf cramps, Neck pains, Tingling in hands and feet, vagotonous regulation in the nervous system, heart rhythm disturbances, cramps of organs and vessels	Peanut butter, soy beans, cocoa, nuts, natural rice, oatflakes, wholemeal bread, peas, milk chocolate, fish, milk, mineral water with magnesium. As medicine: 0.3-0.5 g/day of magnesium preparations
Potassium Serum concentration: < 3.5 mmol/l	Weak muscles, decreased reflexes, diarrhoea, lack of desire to train, fatigue, heart rhythm disturbances. (Numerous medicines influence potassium ingestion.)	Meat extract, legumes, dried fruit, grains, potatoes, bananas, tomatos, meat. As medicine: Potassium magnesium aspartate.
Iron Serum concentration: Ferritin (F): < 30 μg/l (M): < 40 μg/l Iron (F): < 60 μg/dl (< 11 μmol/l) (M): < 80 μg/dl (< 14 μmol/l) Haemoglobin (F): < 12 g/dl (< 7.4 mmol/l) (M): < 13 g/dl (< 8.1 mmol/l)	Fatigue, feeling worn out, lack of desire to train, anaemia (ensure preparations are digestible)	Liver, kidneys, legumes, cocoa, wholemeal bread, liver pâté, spinach, nuts, chocolate, meat (red). As medicine: 10-40 mg/day
Zinc Serum Concentration: < 12 mmol/l	Disturbances of taste and smell, lack of appetite, weight loss, skin changes, fatigue, increased susceptibility to infections	Cheese, full cream milk, meat, oysters, chicken eggs. Because of phytate, poor processing of zinc from legumes and grain.

Table 39

On average 9 to 15 g/day are ingested, mostly in unrecognised forms. The desirable amount is 5 to 8 g/day. Fluids drunk **during exercise** should always contain salt in **isotonic solutions** (0.5 to 1.0 g/l of NaCl). Drinks **containing salt are absorbed more quickly**. There is a delay in the absorption of pure tap water (without salt) in the intestines.

In sport additional salt ingestion is only necessary during extreme heat endurance loads (Ironman, 100 km runs etc.). For usual exercise of several hours additional salt ingestion is not needed. In normal outside temperatures (< 20°C) ingestion of salt concentrates during endurance exercise or other sporting activity is unnecessary. Highly concentrated salt ingestion can cause gastro-intestinal problems. Quantities of salt above 1.5 g/l must first be diluted in the intestines before they can be absorbed.

Trained bodies protect themselves against high loss of salt by ensuring sweat contains less salt. As a result of adaptation the extended **sweat glands** of **competitive athletes** cause **increased mineral reabsorption**. Thus a litre of sweat of an untrained person contains 3.5 g of salt but of a trained person only 1.6g. Therefore an untrained person loses as much salt in 3 litres as a trained person in 5 litres. During extreme endurance exercise about 10-15 litres can be sweated per day; this leads to a salt deficit of about 16-24 g/day. To secure performance capacity and important bodily functions the **blood sodium concentration** must not be allowed to drop below 130 mmol/l. Extremely low blood sodium concentrations of 120 to 125 mmol/l have been measured in Hawaii long triathlon athletes; in this condition some of the athletes had to be hospitalised because of an acute danger of the beginning of brain oedema with loss of orientation.

In situations of extreme sweat loss, dosed salt ingestion during load is necessary for the maintenance of important bodily functions. The dosage should be individually appropriate and can vary from 0.5 to 1.2 g/l of NaCl depending on the salt content of the drink. Drinking too much tap water without salt causes **"water poisoning"** during slow runs (NOACKES, 1992).

Muscle cramps during long duration loads are not caused by salt deficiency alone and they cannot be stopped by increasing salt ingestion either. Electrolyte problems in the case of muscle cramps are very complex and linked to a simultaneous local magnesium and calcium deficit.

After major salt loss, via sweat or persistent diarrhoea, it is recommended to eat foods containing salt (salt, pretzels, fish, milk or meat). In such cases preference for foods with salt occurs instinctively.

Potassium

Potassium is required for the securing of **cell membrane stability, nerve impulse transmission, muscle contraction** and for maintaining **blood pressure regulation**. Potassium is involved in the transport processes in **carbohydrate, protein and fat metabolism**. It is of central importance in the development of coenzyme A and acetyl coenzyme A. Oxidative metabolism in the mitochondria is dependent on the potassium content of the cell nucleus. **Potassium** storage **is predominantly intracellular**; its concentration there is 40 times higher than outside the cell. Intracellular K^+ concentration is 155 mmol/l (150-160 mmol/l) and extracellular concentration is 4 mmol/l (3.5-4.5 mmol/l). In the cell membrane there are eight **potassium channels**. When the channels are open, potassium ions always move in the direction of lower concentration. An energy-dependent pump mechanism restores equilibrium between the ion concentrations of the inner and outer cell spaces. The distribution of potassium and sodium on both sides of the cell membrane is maintained by **membrane potential** of the cell. If there is a lack of energy or oxygen the membrane potential can be destabilised. ATP serves as fuel to maintain functioning of the potassium-sodium pump in the cell membrane, provided by a specific sodium-potassium ATPase. The cofactor for activating this enzyme is magnesium. A further pumping mechanism exchanges sodium ions with calcium ions. Whereas the sodium and potassium ions mainly maintain **cell membrane potential**, the magnesium and calcium ions present serve more as steering and controlling mechanisms. This brief description is intended to show that there is a close interaction of the electrolytes in maintaining membrane stability, especially during sporting load, and that a deficiency of one mineral can lead to complex functional disruptions.

In untrained young men the **potassium stores** of the body contain 140-150 g and those of women 90-120 g. The stores of sport practitioners grow larger. In sprinters **total body potassium** is 166 g (2.37 g/kg), in middle distance runners it is 159 g (2.27 g/kg) and in long distance runners it is 151 g (2.36 g/kg) according to GRUBE (1993); in untrained persons on the other hand it is 142 g (1.94 g/kg).

Minimum ingestion of potassium should be 2-3g/day (see Tables 38 and 39). Sport practitioners need 3-4 g daily. These requirements can be fulfilled with a balanced mixed diet including plenty of fruit. There are no problems with potassium supply from a vegetarian diet. In healthy persons 80-95% of potassium ingested with food is excreted in urine.

The main **potassium sources** in food are citrus fruits, bananas, tomatoes and fruit. On a vegetarian diet up to 10 g/day of potassium can be ingested. **Excretion** takes place via urine, stool and sweat. In comparison to sodium there is minimal loss of potassium through sweat, namely 0.1-0.2 g/l. For sport practitioners potassium is necessary for glycogen storage. Therefore they always have a higher total potassium stock than untrained persons with less muscles and glycogen. When larger quantities of carbohydrates are ingested during training, protein and potassium requirements rise. High muscle use leading to structural damage, such as downhills and low jumps lead to an increased loss of potassium and thus of potassium requirements. Further losses of potassium are possible if the body exchanges potassium for sodium. In a dehydrated state, both in sweat and urine, potassium is increasingly excreted instead of sodium. The **potassium blood level** is 3.8-5.5 mmol/l and changes little after exercise. During exercise, as a result of the breakdown of **muscle glycogen**, more potassium is released which is available functionally and temporarily compensates losses. In body building sometimes more potassium is ingested before a competition. By displacing sodium it leads to dehydration and makes the skin “thinner“. When the subcutaneous tissue is dehydrated the muscle structures become more visible.

Magnesium

Magnesium is an indispensible mineral and is stored in the body in quantities of 24-28 g (584-681 mmol/l). 60% of this is in the **skeleton**, where, however, it is difficult to access. A further 39% is stored **intracellularly**, mainly in the **muscles**. In the serum and in extracellular fluids only 1% of the magnesium is found.

Normal s**erum concentration is 0.75 to 1.1 mmol/l.** In the erythrocytes there are 2.5 mmol/l of magnesium. The great significance of magnesium is explained by the fact that it is a **component** of about **300 enzymes**. Thus magnesium is necessary for the **provision and transmission of energy, signal transmission** in muscle concentration, the **relaxing of muscles, the supply of blood, hormonal activity** and other important functions (Tables 38 and 39). Magnesium is indispensable for the **activation** of the muscle contraction enzyme adenosine triphosphatase **(ATPase)**. The permeability of the **cell membranes (cell permeability)** is dependent on magnesium; it increases when there is a magnesium deficit.

Magnesium deficiency leads to a **decrease in the density of the sodium-potassium pump** in the cell membrane, decreases **ATPase activity** and thus increases **cell membrane permeability**. As a result of these disruptions there is a loss of cellular potassium and depositing of Na^{+} and Ca^{++} in the cell nucleus.

Each day the body loses magnesium through **sweat** and even more through **urine**. At 2 l sweat loss, which is normal in a summer training session, there is magnesium excretion of 9 mmol/l (18 mmol altogether). A decrease in magnesium concentration in the blood after long endurance loads is normal. The reason magnesium loss after long duration sporting loads is so high is that the **larger quantity** is excreted through the **kidneys**. Even after a marathon the body loses an additional 2.5 g of magnesium in a week through urine, over and above normal excretion of 140 mg/day. Total loss of magnesium after strenuous endurance exercise of several hours is an average of 3.5 g/week.

A magnesium **deficiency** can be recognised by determining the **blood concentration**. If concentration at rest goes below **0.75 mmol/l** in the blood plasma, for competitive athletes additional ingestion is recommended. Magnesium deficiency due to food leads to a decrease in blood magnesium (hypomagnesaemia), and at the same time to a potassium deficit in the muscle cells. A deficiency of intracellular magnesium does not occur for another two months.

In competitive sport a magnesium deficiency is always possible because, in addition to magnesium loss through urine and sweat, consuming foods and beverages low in magnesium is favourable to a deficiency. In addition to determining magnesium in the blood, the quantity excreted in urine over 24 hours can be measured. Normally 1 mmol/l of magnesium is excreted in 24/h, less when there is a deficiency.

Recommended additional **ingestion of magnesium** during competitive sport training is **200 to 300 mg/day**; this can be increased to 500 mg/day. It is advantageous to combine **magnesium ingestion with simultaneous ingestion of potassium**. Table 40 contains details of foods with a high magnesium content.

Symptoms of magnesium **deficiency** can be quivering muscles, calf cramps, nervousness, fatigue or decreasing performance capacity. If a magnesium deficit is suspected, the blood concentration of magnesium should be checked. In training control an increased heart rate at rest and/or during load can point to a functional magnesium deficiency. Before great training loads and competitions, especially in heat, as a precautionary measure the magnesium pool should be stabilised through additional ingestion of 200 mg per day over several days. Magnesium preparations to which aspartate has been added show better absorption in the small intestine. Drinking magnesium rich mineral water (over 100 mg/l) can help ensure an adequate supply of magnesium.

Magnesium Content of Foodstuffs*

Food	Magnesium mg/100 g (mmol/l)
Cocoa powder	414 (17,03)
Wheat germ	336 (13,81)
Brewer's yeast	231 (9,50)
Soy flour	235 (9,66)
Peanuts	182 (7,48)
Almonds (sweet)	170 (6,88)
Hazelnuts	156 (6,41)
Oatflakes	139 (5,71)
Beans (white)	132 (5,43)
Walnuts	129 (5,71)
Rice (unpolished)	119 (4,89)
Peas (shelled)	116 (4,77)
Chocolate	104 (4,27)
Lentils	77 (3,16)
Crispbread	68 (2,79)
Pasta (noodles)	67 (2,75)
Raisins (dried)	65 (2,67)
Herring fillet (in tomato sauce)	61 (2,51)
Wholemeal bread (wheat flour)	59 (2,82)
Wholemeal bread (rye flour)	45 (1,85)
Kohlrabi	43 (1,76)
Bananas	36 (1,48)
Potatoes	25 (1,03)
Beans (green)	25 (1,03)
Mineral water	20 - 160 mg/l

Table 40: Sea salts contain a high proportion of magnesium.
** Figures from HOLTMEIER (1995)*

Calcium

Calcium is an important **enzyme activator** and is also functionally necessary for **neuromuscular signal transmission, cell membrane permeability, energy release and blood coagulation.** Calcium is the main component of the bones and teeth as well as being important for their strength.

The body's **calcium stores** contain 1,000 g; 98% of which is found in the bones. Serum contains 2.3-2.7 mmol/l or 90-108 mg/l. After sporting loads there are uncharacteristic changes to calcium concentration in the blood. Calcium requirements are met by the stores in the blood. Normally sport does not lead to a calcium deficiency. Nevertheless demineralisation of bone structures is often found in young female athletes, encouraged by oestrogen deficiency. This demineralisation is a preliminary stage of osteoporosis. In competitive sport this bone development disturbance can be expressed in the form of unexpected **"stress fractures" or fatigue breaks** in bones loaded to a great degree by tractive muscle forces. Hormone deficiency (usually oestrogens) and insufficient calcium ingestion were recognised as the cause. As male athletes are also affected by stress fractures, the causes must be complex. High **sporting load stress**, especially in young women, leads to peripheral hormone deficiency resulting from hypothalamic dysfunction. This central disturbance influences menstruation regularity and in extreme cases interrupts menstruation **(amenorrhoea)**. Medical specialists recommend therapeutic ingestion of oestrogens, progesterone and calcium to women affected this way.

Daily calcium requirements are 900-1,200 mg. Milk, dairy products, cheese, hazelnuts, legumes and vegetables are rich in calcium. Thus daily requirements would be covered by a litre of milk or 150 g of hard cheese. Recently the possibility has been discussed that the high protein of milk may hinder calcium ingestion. When calcium is absorbed, interactions through phytate, oxalate and phosphate (meat, Coca-Cola) are to be expected. This impediment to absorption, the possibility of which cannot always be completely excluded, justifies taking additional calcium.

In addition to giving preference to foods containing calcium, to be on the safe side calcium preparations can be taken. For highly loaded young athletes, especially in the growth phase, this is a practical way of ensuring the calcium balance is maintained. Calcium requirements increase with increasing energy intake (calorie consumption) of sport practitioners (see Table 38). If there is a calcium deficiency over several months, a lowering of bone density is probable. Unexpected occurrence of stress fractures in competitive sport may be an expression of calcium deficiency.

Information on **bone density**, serum calcium level and calcium excretion, in athletes is not uniform because the type of sport, the training state and current diet are rarely taken into consideration. After long duration loads there are no conspicuous changes in serum calcium concentration. Calcium is also lost in **sweat**. The calcium content of sweat varies greatly, however, and is given as 5 - 50 mg/l. The DGE recommendation of 1,200 mg/day of calcium ingestion is not enough for 14% of athletes (ROKITZKI, 1994). In game and strength sports, **calcium requirements** are considerably higher, namely 1.5 - 2.5 g/day.

Iron

The trace element iron is a component of **oxygen transporting compounds** in **haemoglobin, myoglobin and enzymes**. The body's stock of iron is 3-5 g, of which 70% is bound in the red blood colouring (haemoglobin) as haemic iron. The **functional iron reserves** are 2.3 g in the haemoglobin, 0.32g in the myoglobin and 0.18g in enzymes containing iron (cytochrome, catalase, peroxidase). Functional iron makes up 12% of the body's stocks. The other 18% is deposited in the **iron store complexes,** e.g. 700 mg in ferritin and 300 mg in haemosiderine. Only 0.1% of the iron is transported in the blood in a special protein, **transferrin**, and is highly utilised because of the low quantity. The **liver** is the **largest iron store**. The most **reliable diagnostic parameter for the state of the iron store is the ferritin concentration in the blood** (see Table 39). The iron concentration in the blood is 0.6-1.45 mg/l (10.7-26 μmol/l) in women and 0.8-1.68 mg/l (14.3-30 μmol/l) in men. The blood iron level is not enough for the diagnosis of an iron deficiency. In sport the ferritin concentration is more meaningful. The **normal range of ferritin** fluctuates between 30-400 μg/l in male athletes and 30-150 μg/l in female athletes. When the ferritin concentration drops below 30 μg/l a beginning of iron deficiency is probable. Sport practitioners should aim for an average ferritin concentration of 30-150 μg/l (HOFFMANN, 1995).

Ferritin is the most frequently used indicator for early recognition of **iron deficiencies**; only then follow transferrin and transferrin saturation. The isolated determination of serum iron only has limited meaningfulness. When evaluating the haemoglobin concentration, the increase in plasma volume through training must be taken into consideration, especially with endurance trained athletes. This

hypervolaemia is the cause of **"pseudoanaemia"** amongst competitive athletes. This adaptive **diluting effect of the blood** can be recognised in the decreasing haematocrit (increased number of fluid blood components). When the haematocrit is normal the haemoglobin concentration in competitive sport must not drop below 13 g/dl in men and 12 g/dl in women (see Table 39).

After longer sporting loads the ferritin concentration can rise by 20-30% for several days so that a deficit in the stores is concealed by the high utilisation or as an acute phase protein reaction.

Female endurance athletes and strict vegetarians are among the people who have systemic iron deficits. The main reason for the iron deficit in competitive athletes is the **mechanical destruction of the erythrocytes** in the areas coming into contact with sports equipment or in the soles of the feet in running. Many **runners tend to have low iron levels** so that regular ferritin and transferritin checks are necessary. When there are signs of a deficiency it is necessary to

Foods with High Iron Content

Food	**Iron (mg/100 g)**
Liver (pig)	22
Yeast (dry)	17
Cocoa	12
Soy beans, lentils, white beans (legumes)	7 - 9
Oatflakes, wheat germ	5 - 8
Parsley, spinach	4 - 8
Almonds, hazelnuts, sunflower seeds	3 - 7
Apricots, figs (dried)	3 - 5
Whole grain products	3 - 4
Beef, veal	3
Chocolate	3

Table 41

supplement with iron preparations over longer periods. After daily ingestion of 100 mg of bivalent iron over six weeks a ferritin increase of 15 μg/l can be expected. Digestibility of the individual preparations varies which in practice limits fast refilling of the stores.

The special protein that protects the body against loss of iron is haptoglobin. **Haptoglobin** binds the haemoglobin released from destroyed erythrocytes. If this does not happen completely, large quantities of iron are lost in urine. In competitive athletes the haptoglobin concentration must not drop below 0.5 g/l in the serum. If the haptoglobin values are too low, the body's own production can no longer keep up with the quantity necessary to intercept the haemoglobin released by the erythrocytes.

Other **areas** of **iron loss** are **menstruation** and excretion via the stool as well as **sweat**. Iron losses through sweat fluctuate; if there are 2 l of sweat, an iron loss of 0.3 mg is to be expected. In addition, small **bleeding spots** in the **gastro-intestinal tract** and in the **bladder** during extreme loads lead to iron losses. If an athlete has changes in the blood picture, or a drop in haemoglobin to below 12 g/l, there exists a longer iron deficiency which needs medical treatment.

Training athletes lose about 2 mg of iron per day. Competitive athletes should therefore always ensure regular iron ingestion through food. **Liver, meat, whole grain products, legumes and fruits** are **especially rich in iron** (Table 41). Of the meat varieties, red muscle meat is the main haem iron source. Other sources are cocoa, wheat germ, whole grain products and chocolate.

On average only 10% of the iron in food is absorbed, that means ingestion of about 29 mg/day is necessary in order to compensate for loss. Eating habits can influence iron absorption in the intestine. Of **vegetarian food** only 3-8% of the iron contained is absorbed and when meat is eaten 15-22%. The high level of **phytic acid** in plants **impedes iron absorption** through the development of complexes beforehand in the intestines. In the presence of meat, iron absorption from plants can be increased considerably. **Vitamin C also improves iron ingestion** in the intestines. In practice, through simultaneous drinking of orange juice (fruit juices), or ingestion of vitamin C preparations, absorption of iron can be doubled or quadrupled. In contrast, drinking coffee, tea, Coca-Cola or very fibrous foods has the effect of **reducing iron ingestion**. If there is low iron ingestion over a long period the iron stores are exhausted after 5-8 months. Refilling takes a correspondingly long time. Vegetarians more frequently have iron deficiencies than those eating a mixed diet.

If there is a drop in the serum iron level during infectious illnesses it is not necessary to supplement. Additional iron ingestion might even disrupt defence against the infection.

In **competitive sport** the following situations can disturb the state of equilibrium in **iron metabolism**:

- One-sided vegetarian diet (vegetarianism)
- Insufficient ingestion via the intestines (absorption inhibitors)
- Iron losses through anaerobic and long aerobic metabolism load stress in the muscles
- Mechanical hindrance in the gastro-intestinal tract (mucous membranes) and bladder function
- Mechanical traumatisation of the soles of the feet (hard ground in long distance running)
- Excessive sweating (0.1-0.2 g/l) and
- Menstruation (15-30 mg/cycle).

Additional iron ingestion cannot increase sporting performance capacity. A **deficiency**, however, will lead to a **drop in performance**. Competitive athletes, in particular **women**, should regularly have their **iron status checked**. Indications of a deficiency are: unaccustomed fatigue, premature exhaustion, performance stagnation, increased breathing during load as well as frequent infections.

Zinc

Zinc is a **component of many enzymes**. In the body 1.3 to 2 g are stored, mainly in the bones and muscles. The stores are stable and not immediately available for metabolic processes. Only the quantity of zinc in the blood of 4-7.5 mg/l (61-114 μmol/l), 90% of which is deposited in the **erythrocytes**, can be used directly for metabolic processes. The great physiological significance of zinc in competitive sport moves to the centre of attention for the following reasons:

- Zinc aids protein synthesis (anabolic protein processing)
- Zinc impedes protein breakdown
- Zinc reduces cell damage during extreme exercise and
- Zinc supports the reactivity of the immune system.

Serum concentration is 9-18 μmol/l (0.6-1.2 mg/l), whereby competitive athletes are at the upper limit with an average of 17.5 μmol/l. After marathons the zinc concentration in the blood rises by 23% until the third day (results of own research). The cause is to be seen in the mechanical **destruction of the erythrocytes**, similar to the release of iron and magnesium. In addition there is a zinc loss of 0.1-0.3 mg/l through sweat.

Competitive athletes on a varied diet normally do not have a zinc deficiency. The zinc ingestion recommended by the DGE of 10-15 mg/day is covered (see Table 38). With increasing energy ingestion competitive athletes also ingest more zinc. At 4,000 kcal the figure is 15-20 mg of zinc daily. Meat, liver, sea fish, milk and eggs have the highest zinc content. Refined foods contain very little zinc. As with iron and magnesium, a one-sided vegetarian diet can lead to a zinc deficiency. A **vegetarian diet**, rich in fibres and phytates, **impedes zinc absorption** in the intestines. This does not mean that all vegetarians practising sport have zinc deficiencies. A normal zinc concentration is necessary to ensure ability to handle load, to prevent muscle cramps and to support the function of the immune system. A **zinc deficit** slows lactate breakdown and impedes protein synthesis; intended muscle hypertrophy in strength training (body building) fails to materialise. For these reasons **higher zinc ingestion** is recommended for endurance and strength athletes, namely 20-30 mg/day.

Trace Elements

The trace elements are a **sub-group of the minerals** which are only found and ingested in the human body in **small quantities**. The normal fields can fluctuate considerably; the concentrations in the various tissues vary greatly. Trace element samples tell us something about the source (e.g. hair, nails, sweat, urine), but not about the content in the body as a whole or in the needy tissues. Trace elements can be frequently ingested in mega-dosages. **Mega-dosages** means exceeding the DGE or RDA recommendations by ten times. Trace elements are divided into essential, semi-essential and those without a known physiological effect as yet.

The group of **absolutely necessary (essential) trace elements** includes:
Chromium, iron, cobalt, copper, manganese, molybdenum, nickel, selenium, iodine, fluoride, silicon, vanadium, zinc and tin (Fig. 26).

Those not certainly indispensable, or **non-essential** trace elements are:
Aluminium, antimony, arsenic, gold, barium, beryllium, lead, boron, bismuth, caesium, cadmium, mercury, silver, strontium, lithium, rubidium, platinum, thallium, tellurium, titanium and inert gases. Some non-essential heavy metals can be poisonous in high doses (mercury, strontium, thallium among others).
Only **selected trace elements** with a certain significance for the securing of ability to handle sporting load are dealt with here.

Differentiation between minerals and trace elements is done according to daily requirements. Thus those **minerals for which daily requirements are below 100 mg/day are included among the trace elements.**

Copper

Copper is a **necessary trace element** for the body; a copper deficiency leads to impairments to health, disruptions to tissue development and restrictions of enzyme activity. Copper is a component of 16 essential metalloproteins necessary for the development of **connective tissue**, the functions of the **central nervous system** and **blood production**. An absolute copper deficit results in death and too much causes poisoning. The copper transporting protein in the blood, **coeruloplasmin**, is involved in immunological defence; it is a **protein of the acute phase**.

80-150 mg (1.5-2.35 mmol/l) of copper are stored in the body. The serum concentration of copper is 0.70-1.40 mg/l (11-22 μmol/l) in men and 0.85-1.55 mg/l (13-24 μmol/l) in women. The supply of copper required daily is given as 1.5-3 mg/day and can be covered with a normal mixed diet. It is possible that **competitive athletes** need **more copper** because increased restructuring in muscle and tendon tissue as a result of training load shows an increased need for copper. Of the copper ingested, about 25-30% is excreted again with **sweat**, where the quantities vary greatly (ANDERSON, 1991). The main place of excretion of copper, however, is the gallbladder. Foods particularly rich in copper are liver, fish, nuts, cocoa and legumes. Meat, milk, root vegetables and pasta on the other hand contain little copper.

Overdoses of copper can disrupt zinc reactions in enzymatic processes. Plentiful iron and zinc ingestion results in decreased copper absorption. Excess ingestion of copper causes diarrhoea and cramps, and in extreme cases haemolytic anaemia.

Selenium

Selenium is an **essential trace element** and 20-100 μg must be ingested daily. As a component of the principally **antioxidative enzyme glutathione peroxidase** it protects the erythrocytes against destruction (haemolysis) by oxygen radicals. Recently selenium was also found to be a component of a thyroid hormone producing enzyme (iodothyromine-5'-deiodase). This means that a deficit of the vital enzyme thyroxine-deiodase in the case of a selenium deficiency leads to typical **iodine deficiency symptoms**. An iodine deficit means cold sensitivity, low blood pressure, body fattiness, hair and skin changes as well as goitre. During fatty acid breakdown **malondialdehyde** arises, an important indicator of the performance capacity of the antioxidative protective system. **Glutathione peroxidase**, which contains selenium and is highly antioxidative, reacts **synergetically with vitamin E**, i.e. together they neutralise the free radicals arising during fatty acid breakdown.

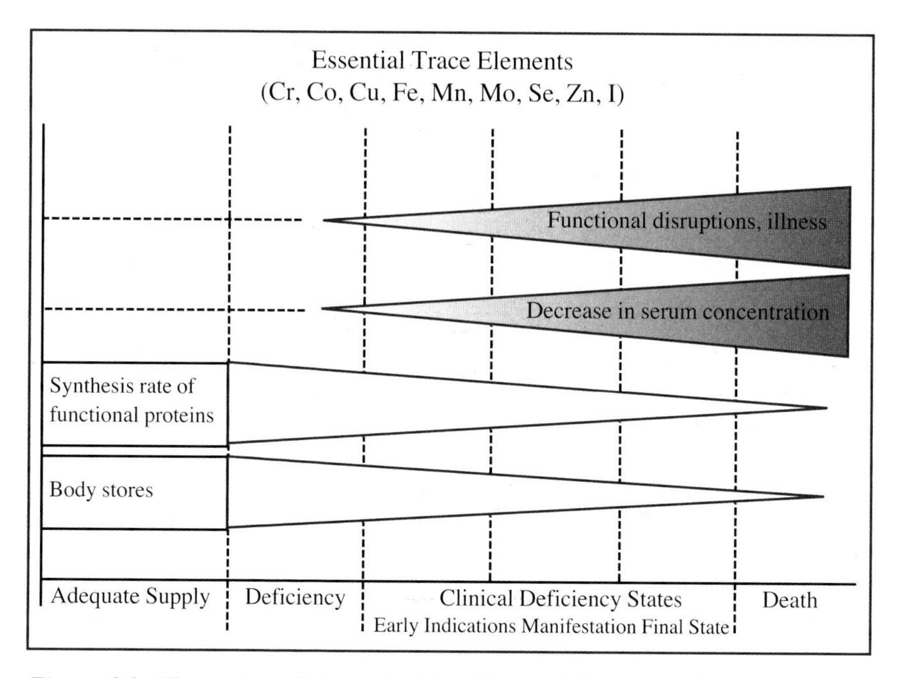

Figure 26: Illustration of the principle effects of the essential trace elements (chromium, cobalt, copper, iron, manganese, molybdenum, selenium, zinc and iodine) when there is a deficiency

The body's stocks of selenium are found mainly in the kidneys, liver, skeletal muscle and the erythrocytes, and amount to 10-15 mg. In Germany the serum concentration fluctuates between 50-120 μg/l (0.6-1.5 μmol/l). The DGE recommends daily ingestion of 20-100 g of selenium, but experts consider the lower amount to be too low.

It is **difficult to establish the normal value** because the selenium content of foods depends on the region of production (selenium content of the soil), environmental pollution with heavy metals. and eating habits. The selenium content of the erythrocytes provides information on long term supply, while measuring the selenium content of the hair or nails is less certain.

The supply of selenium comes from sea fish, meat, liver, grain products, yeast and nuts. Deficits are known in regions with selenium impoverished soils (China). Alcoholics can have a selenium deficiency. In cases of selenium deficiency deformities in the heart and skeletal muscles (myopathy), as well as increased haemolysis (Keshan sickness in China) occur.

Selenium requirements in sport cannot be established for certain. They are assumed to be **50-100 μg/day**, however, in the USA 1 μg/kg of body weight. Sometimes considerably larger quantities are recommended, namely 150-250 μg/day (SCHRANZER, 1997). Preference should be given to ingestion of organic yeasts containing selenium in order to avoid possible indigestibility of inorganic selenium (ZIEGLER, 1997). Ingestion of 1 mg of selenium yeast is the equivalent of 1 μg of inorganic selenium. The significance of selenium for production of the antioxidative enzyme glutathione peroxidase in the cellular immune system, for the heart muscle and blood vessel endothelium, reproduction and the aging of tissue justify its increased availability and ingestion by competitive athletes. Germany is considered a **selenium deficient** country.

Chromium

Chromium is an **essential trace element**. Daily requirements are assumed to be 0.01 mg (0.2 μmol). Chromium is known as a cofactor of insulin and forms a complex with it. Many diabetics have a chromium deficiency. Chromium is involved in **carbohydrate and fat metabolism** and increases glucose tolerance when amply available. Chromium aids the **development of glycogen stores**. Through increased release of fatty acids it is thought to be suitable for **losing weight**.

Chromium is necessary for the activation of the effects of insulin. In serum chromium is transported with transferrin and albumin. The **serum concentration** is about 0.5 μg/l (9.7 nmol/l), but it is not a meaningful indicator of chromium supply. A **deficiency** results in reduced glucose tolerance and increased insulin, leading to hypoglycaemia and damage to peripheral nerves. In animal experiments it was found that if there is a chromium deficiency the glycogen stores decrease, and if it is supplied in great quantities these enlarge again. Increased chromium ingestion aids the incorporation of amino acids in muscle tissue and increases fat utilisation (CAMPBELL/ANDERSON, 1987; ANDERSON, 1991). A number of double blind studies of ingestion of 200 mg/day of **chromium picolinate** over several weeks did not show any effects on muscle strength, fat metabolism and the decrease in body fat (HALLMARK et al., 1996). If chromium is ingested in large quantities liver damage can occur.

Symptoms of Mineral Loss

Mineral deficiencies are not easily recognisable. Certain indications are muscle cramps during longer loads, disruptions to heart rhythm, muscle fatigue, longer regeneration. If there is suspicion of a deficit, first of all the mineral concentration in the blood should be determined.

In **competitive training** there are often **deficiencies of magnesium, iron, calcium, zinc and selenium. Muscle cramps** occur frequently when the **blood magnesium level** drops below 0.7 mmol/l, coupled with a local calcium deficit in the muscles. Muscle cramps from a low selenium-magnesium level are not always recognisable, so magnesium consumption is recommended even when blood concentration is normal. **Heart rhythm disturbances** in connection with mineral deficits can be influenced by taking potassium magnesium aspartate preparations. Especially when there is a **great loss of sweat**, as in summer endurance training of over 12 h/week, **multimineral preparations** should be taken regularly as a preventative measure.

Side Effects of Vitamin and Mineral Ingestion

The **additional ingestion of vitamins and minerals in the physiologically recommended dosages is not damaging to health.** Even in the case of plentiful ingestion of these active substances there is **no danger of addiction**. **No overdoses** of the water soluble vitamins (B_1, B_2, B_6, B_{12}, nicotinamide, pantothenic acid, biotin, folic acid among others) need be feared because excesses ingested are excreted through the kidneys. The fat soluble **vitamin E** can also be taken in very high doses **(over 400 mg/day)** and causes no damage. With vitamins A and D overdoses are possible so medical advice may be appropriate.

Of the minerals, **iron preparations** ingested in large quantities can cause **problems** in the **gastro-intestinal tract**. If treatment becomes necessary a different preparation should be used and one found that is digestible for the individual. The ingestion of too **much salt** can also disturb the gastro-intestinal functions considerably because body water for dilution must flow into the intestines. In the past when salt consumption at mass running events was in high doses, there were stomach cramps. Today it has become accepted practice to add **0.5 to 1.0 g/l of salt to fluids consumed**; this way absorption is ensured. High dosages of salt are not dangerous to health. Problems in the intestines continue until osmotic compensation is complete. The minimum toxic dosage of **salt is about 100 g**, a quantity which cannot be ingested in sport.

Other minerals could also be ingested in high dosages without causing damage to health. With potassium, calcium and phosphorus the toxic dosage if taken all at once is 12 g, with magnesium it is 6 g, with zinc 0.5 g and with iron over 100 mg. If the body is completely supplied with minerals, **through self-regulation, absorption through the intestines decreases**, i.e. the excess ingested is excreted.

8 Active Substances and Performance Capacity

8.1 Amino Acids

The proteins or rather their constituents, the amino acids, must be ingested in sufficient quantities to prevent a negative nitrogen balance occurring. The protein density in food, and also their biological valency, is of great significance for adolescents, sport practitioners and also older people. The negative consequences of a deficiency of certain amino acids were recognised through slimming diets where e.g. taking gelatine preparations led to a deficit of sulphurous amino acids. The amino acid preparations available on the market are usually made from milk, whey, soy beans, fish or grain (Table 42).

The amino acids **ornithine** and **arginine** attracted medical interest at an early stage after it was proved that through their involvement in the production of urea they increase the detoxication of blood ammoniac, which occurs in greater amounts in sport. These two amino acids increase the supply of fatty acids to the muscles, promote cell structure restoration, influence immunity positively and greatly activate the growth hormone (ALBINA et al. 1988; BARBUL, 1985; SAITO et al. 1987). Thus in a completely natural way they have an effect, which counters protein catabolism, as a result of exercise.

The major prerequisite for the anticatabolic effect or the reduction of catabolism is that arginine and ornithine must be supplied together in considerably higher doses than is possible in normal food.

The amino acids ingested with normal food only have a very untargeted effect.

The effect of the two amino acids **arginine** and **ornithine** on the **growth hormone** (GRH) was discovered when infusions were carried out for clinical parenteral feeding. Oral ingestion of arginine and ornithine had little effect or could not be scientifically proved because of underdosing and ingestion at the wrong time. After it was established that an empty stomach was favourable to absorption and that the dosage must be higher, interest in these two amino acids in competitive sport suddenly increased.

The secretion of GRH is increased after 90 min if 170 mg/kg of ornithine is taken orally (BUCCI et al. 1990). The effect on the secretion of the growth hormone is considerably greater when ornithine and arginine are taken together. The effect is not completely clear in a physiological sense, but has been clearly proved in experiments. The release of the growth hormone occurs at night during the body's resting phase.

The main reason for deliberate **ingestion of amino acids** is to secure stable muscle **regeneration** in phases of great load and in competition series, as well as to promote **muscle development** after strength training in competitive sport.

The ingestion of amino acid concentrates, and individual substance components is an alternative for competitive athletes if they do not want to risk a starting ban of up to two years for the unpermitted taking of anabolic steroids.

Overdose problems from the taking of amino acids by healthy persons are unknown. People with kidney problems are the exception as they cannot sufficiently detoxicate the increased amounts of urea which occur as a result of amino acid catabolism.

The increased amino acid breakdown which occurs in competitive training is recognisable from the increased **serum urea level**. Whereas in training with effective stimuli the serum urea concentration levels off at values between 5 and 7 mmol/l (in women 1 mmol/l less); in cases of overloading in training urea can rise to 9-11 mmol/l (Fig. 27). After one-off extreme loads (100 km run, 24 hour run) the high rate of protein breakdown leads to an increase in serum urea to up to 15 mmol/l. It should be noted that one-off ingestion of larger amounts of protein (e.g. 200 g steak, 300 g cheese or amino acid concentrate) additionally increases serum urea by 1-2 mmol/l.

The body cannot change all ingested amino acids in metabolism. This has resulted in the amino acids being divided into **essential, semi-essential and non-essential amino acids.**

Essential amino acids are those which the body cannot produce and is thus dependent on them being constantly supplied. Semi-essential amino acids are produced by the body to a limited extent. Non-essential amino acids can be transformed into each other in metabolic processes.

- The **essential amino acids** are:
  Leucine, isoleucine, valine (branch chain amino acids), lysine, methionine, phenylalanine, threonine and tryptophan.
- The **semi-essential amino acids** are:
  Histidine, arginine, cyteine (cystine) and tyrosine.
- **Nonessential amino acids** are:
  Alanine, asparagine (asparagic acid), glutamine (glutamic acid/glutamate), glycine, proline (hydroxy proline), serine and ornithine.

From a physiological point of view the following **amino acid effects** are known:

- **Support of Hormone Production**
  Ingestion of arginine and ornithine increase the spasmodic release of the growth hormone (GRH), as BUCCI et al. (1990) were able to prove. This provides a reason for additional ingestion of this amino acid in quantities of up to 12 g/day to support muscle growth. The training stimulus in resistance training (weight lifting) evokes a similarly strong activation of GRH by just supplementing with arginine and ornithine in a ratio of two to one (ARNDT, 1994). Increasing arginine intake to 30 g/day did not lead to any undesirable effects. The theoretical views on deliberate supplementation of these amino acids, particularly in weight lifting, vary. Increased ingestion of these amino acids is practised in body building and in maximum strength sports using dietetic foods.

  Ornithine is only ingested in small quantities with the proteins in food. The body produces this amino acid in the liver from arginine. Ornithine stimulates the release of insulin and supports the breakdown to urea of the increased amounts of ammoniac occurring during intensive muscle loading.

  Influencing of the increased pulsatile release of GRH is also possible through a range of other amino acids (cyteine, glycine, histidine, phenylalanine, lysine, tryptophan and tyrosine).

- **Promotion of Gluconeogenesis (glucose production)**
  The branched-chain amino acids valine, leucine and isoleucine are very useful in gluconeogenesis because they provide the nitrogen for the production of pyruvate. They can be ingested in quantities of 1 to 20 g/day. Research by BLOMSTRAND et al. (1991) showed that ingestion of 4-16 g of branched chain amino acids had a glucose stabilising and performance supporting effect during the marathon. At the same time the decrease in glutamine, a fatiguing

Amino Acids in Sport

Amino Acids (AA)	**Effect**	**Effect in Sport**	**Dosage in Sport** (per day)
Asparagine	Reduces ammoniac development	Increase in aerobic endurance performance	3 - 10 g
Tryptophan	Increases release of growth hormones (GRH), increases synthesis of 5-hydroxy tryptamine and serotonin in the brain, promotes sleep	Anabolism, increase in endurance performance	1 - 1.5 g
Branched-chain AA (Valine, leucine, isoleucine)	Promotes gluconeogenesis, impedes entry of tryptophan into the brain	Increase endurance performance, delay fatigue, altitude training easier to handle	2 - 10 g
Arginine	Stimulation of STH	Muscle development, promotion of fat processing	8 - 10 g
Ornithine	Stimulation of STH	With arginine muscle development	8 - 10 g
Glutamine Acid	Regulator AA, especially for NH_3, promotes protein synthesis in the muscle and glycogen production	Promotes regeneration (glycogen production, immune function)	4 - 6 g

Table 42

substrate, was reduced by ingestion of these amino acids. The branched-chain amino acids work best when taken in conjunction with vitamin B_1, biotin and pantothenic acid.

The increased glucocorticoids during long duration exercise activate the enzymes for gluconeogenesis and energy metabolism in the period of protein catabolism (BLOCK/BUSE, 1990). Thus the rise in cortisol during increasingly occurring gluconeogenetic metabolic processes in competitive sport is normal.

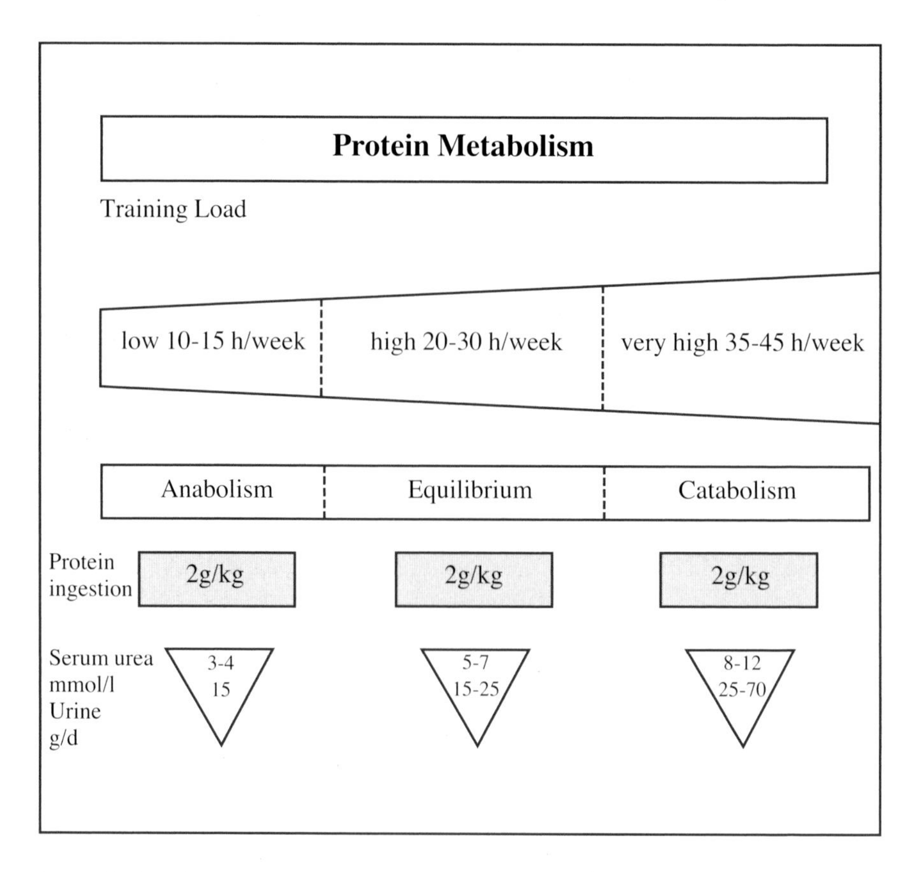

Fig. 27: Relationship of the degree of protein catabolism (protein breakdown) to training load. In cases of major protein catabolism the serum urea level rises considerably, and increased amounts of nitrogen are excreted through the kidneys. In the model illustrated a constantly high supply of protein is assumed.

- **Promotion of Sleep**
  Through increased serotonin development in the brain, tryptophan has the effect of promoting sleep. In this capacity it is recommended as a mild sleeping drug in a dosage of 1-1.5 g/day and should be taken in the evening. In combination with arginine and ornithine it supports muscle anabolism.

- **Increased Muscle Development**
  A number of amino acids have a building-up (anabolic) effect on muscle metabolism. These amino acids include arginine (2-12 g/day), ornithine (2-12 g/day), tryptophan (1-2 g/day), valine, leucine and isoleucine (1.6 g, 2.2 g and 1.6 g/day respectively). Anabolism is already promoted by leucine in amounts of 2.2 g/day through the stimulation of insulin release. Together with GRH, insulin supports the build-up of new tissue structures and promotes muscle development through training. Glutamate, the salt of glutamic acid, has great significance for anabolic metabolism. If the dosages of it contained in ingested amino acid mixtures is too low that reduces the valency of these mixtures.

- **Influence on the Immune System**
  Glutamine can stabilise the defence performance of the immune system. Maintaining the functionality and defence readiness of the immune system is of great significance for the avoidance of overtraining (PARRY-BILLINGS et al., 1992). In the form of glutamic acid, glutamate is a rapidly working excitatory neurotransmitter in the brain. Neurotransmitters serve the transmission of excitation in the synapses of the nerve cells.

- **Promotion of Endurance Performance Capacity and Regeneration**
  Ingestion of 5-10 g of glutamine (as glutamine peptide) has a sparing effect on the glycogen stores and shortens regeneration. By ingesting glutamine during a long lasting load, gluconeogenesis is supported. The branched-chain amino acids (valine, leucine and isoleucine or BCCA) play a key role in supporting regeneration and securing of energy during long lasting exercise. They are an indispensable carbohydrate substitute in situations of carbohydrate deficiency. Carbohydrate deficiency is typical for exercise of several hours. The author's own research has shown that ingestion of 13 g of BCCA counters a decrease of these amino acids in a triple long triathlon and promotes stability of performance capacity. Performance improvements in 40 km timed cycling and in marathons after ingestion of 16 g of BCCA have been documented (HELFER et al. 1995; BLOMSTRAND et al. 1991). BCCA reduces muscular atrophy from walking loads at high altitudes (SCHENA et al. 1992).

There is no uniform opinion about the effect of amino acid supplementation. The free amino acids or their short chain bonds (two to four amino acids are considered short chain peptides) are absorbed more quickly. They are more effective than protein concentrates, certain protein hydrolosates or natural proteins in food. Ingestion of amino acids or short peptides makes itself noticeable in particular through accelerated muscle regeneration. Modern competitive sport with loads of 30 to 50 hours/week needs effective regenerational measures. These become especially necessary when muscles are greatly traumatised by practising a sport, as e.g. during eccentric muscle contraction. Running downhill is one of these unaccustomed forms of muscle use and leads to increased release of creatine kinase from the intracellular space in the blood. In addition, muscle fibres are destroyed and myosine heavy chain fragments are released with a delay from the slow twitch muscle fibres (STF).

In cases of extensive and intensive muscular overuse, today's state of knowledge indicates that deliberate ingestion of additional amino acids is sensible, while requiring expert advice beforehand. Vegetarian athletes in particular should additionally ingest high valency amino acids so they have no disadvantages with regard to muscular adaptation in comparison to "meat eaters".

When ingesting amino acids through diet products the following effects can be expected:

- **Reduction of protein catabolism**
- **Accelerated regeneration**
- **Promotion of muscle development**
- **Protection against overtraining**
- **Stabilisation of the immune system**
- **Promotion of glycogen production**
- **Support of gluconeogenesis**
- **Dietary support of risk groups in sport (see chapter 3.1)**

No overdoses are known amongst athletes who have taken without problems manufactured amino acid mixtures in the biologically effective L form. Academic papers on amino acid supplementation in sport will soon provide further documented findings.

8.2 L-Carnitine

L-carnitine is a **substance produced by the body** from the amino acids methionine and lysine in the liver, testicles and kidneys. The body has a reserve of **L-carnitine** of 20-25 g which is complemented by the daily ingestion of meat, milk and dairy products (Table 43).

The muscles are the largest carnitine store, about 98% of a total of 20 g of carnitine is stored in them. The muscles cannot produce carnitine. The body only synthesises 25% of its carnitine requirements itself. The greatest part must be supplied in food; this applies to an average of 200 mg (100-300 mg) per day. The main source of carnitine is meat. Mutton and beef are the most productive foods containing carnitine, but they would have to be consumed in amounts of 150 to 250 g/day in order to meet requirements.

L-carnitine is a substance necessary for energy metabolism and is mainly needed to transport long chain fatty acids through the mitochondria membrane. The plasma concentration of L-carnitine fluctuates from 40 to 80 μmol/l. Competitive athletes can get into a state of L-carnitine deficiency.

The main causes of relative L-carnitine deficiency are the following:
- Increased energy utilisation, linked with an unbalanced diet
- Deficiency of proteins, vitamin B_6, vitamin C and iron
- An emphatic vegetarian diet with a meat ingestion deficit
- Increased renal excretion after exercise which reduces the muscle depots, and
- Reduction of the muscle reserves through insufficient regeneration during uninterrupted training or competition series (running training, cycle tours).

Six factors are necessary for carnitine synthesis: Lysine, methionine, niacin, vitamin B_6, vitamin C and iron (LEIBOVITZ, 1993).

In competitive athletes endogenic L-carnitine production is made more difficult if the supply of vitamin B_6, vitamin C and iron is insufficient for a longer period of time. Dietary analyses show that about 20% of long duration endurance athletes have a deficiency of vitamin B_6 (ROKITZKI et al., 1994). In 5-10% of long distance runners mechanical haemolysis of the erythrocytes leads to an iron deficit; their ferritin is below 20 μg/l. Iron deficiency among endurance athletes is increased by the additional loss through sweat, which contains 140-275 μg/l of iron.

Many competitve athletes prefer a vegetarian diet, and in addition to low exogenic L-carnitine ingestion often have an iron deficit and a deficiency of lysine. With the help of sports medical advice some competitive athletes have managed to change back to a normal mixed diet, or have considerably stabilised their performance capacity through deliberate L-carnitine supplementation (FÖRENBACH et al. 1993). For a long time there was no direct evidence of a muscular deficiency of L-carnitine in competitive athletes. Muscle bioptic comparative studies with athletes supplied with placebo and L-carnitine documented the theoretical assumptions regarding the usefulness of refilling the muscle stores with L-carnitine (ARENAS et al. 1991). Long duration endurance training of runners led to a greater decrease in muscular L-carnitine in comparison to sprinters. Among the runners who received the placebo there was significantly greater total carnitine excretion in the urine, mainly of the acetyl carnitine fraction, in comparison with the other group of those who supplemented 1 g of L-carnitine (ARENAS et al. 1991).When L-carnitine is sufficiently available it simultaneously stimulates aerobic and anaerobic energy metabolism because through the reaction with acetyl coenzyme A more free coenzyme A is released. The boosting of energy utilisation makes L-carnitine interesting for competitive sport (NEWSHOLME, 1983; MARCONI et al. 1985; DRAGAN et al. 1987; FÖHRENBACH et al. 1993 among others). Nevertheless, with additional ingestion of L-carnitine no direct effect of performance improvement is achieved.

Of the biochemical possibilities for influencing muscular performance capacity through L-carnitine ingestion, four levels of effect have been seen as useful:

1. Increase in mitochondrial fat oxidation
2. Increased rate of pryruvate oxidation
3. Increased glycolysis and alactic metabolism
4. Stabilisation of the immune system and the cell membranes.

In numerous studies attempts have been made to document the effect of L-carnitine supplementation on performance capacity. The results were very contradictory.

With additional ingestion of L-carnitine in doses of 1 to 3 g/day over one to two weeks the following effects can be expected:

- Increased utilisation of long chain free fatty acids
- Increased aerobic and anaerobic carbohydrate metabolism
- Increased stability of the cell membrane and thus of immunological defence potential
- Promotion of regeneration after extreme muscle loads, as well as decreased L-carnitine deficiency among vegetarian athletes.

L-carnitine is non-toxic, does not cause addiction and is not a doping substance. Among highly loaded athletes additional L-carnitine ingestion is often common and is felt to be useful (Table 44). Small doses too, from 0.3 to 1 g, are useful in cases of high loads, also within the framework of fitness sport and above all for older athletes. There is no knowledge of a cut-back in the body's own production when L-carnitine ingestion is high. To be on the safe side, however, ingestion breaks should be taken.

L-Carnitine Content of Foods
(from MITCHEL, 1978)

		mg/l
Meat:	Mutton	210.0
	Lamb	80.0
	Beef	60.0
	Pork	30.0
	Rabbit	20.0
	Chicken	7.5
Other foods:	Yeast	2.4
	Milk	2.0
	Egg	0.8
Grain Products:	Bread	0.2
	Wheat germ	1.0
Vegetables:	Potatoes	0.0
	Cabbage	0.0
	Spinach	0.0

Table 43

Indications for L-Carnitine Ingestion in Sport*

- Promotion of fat utilisation
- Increase aerobic energy metabolism
- Reduction of lactate build-up
- Economical of the glycogen stores and low protein breakdown
- Shortening of regeneration time
- Breakdown of "flab" and increased blood fat values, e.g. of "neutral fats" (triglycerides) and low density lipoproteins
- Stabilisation of cell membranes
- Increase in immunological defence
- Improvement of blood supply to muscles
- Protection against overtraining

**) 1 - 2 g/day over several weeks is advisable*

Table 44

8.3 Ubichinon (Vitamin Q)

Ubichinon is a substance closely related to vitamins and structurally similar to vitamins E and K, and is also called coenzyme Q 10. It is produced in the body and also ingested with food. Its physiological effect consists of the promotion of the **transport of electrons** through the cell membrane. As cofactor in the enzyme complex of the **electron transport chain** it supports ATP production. Furthermore, ubichinon has significant **antioxidative properties**.

Average plasma concentration is 1 mg/l. The more slow twitch muscle fibres (STF) an athlete has, the higher the ubichinon content (KARLSSON, 1997).
It seems there is an inverse relationship between the plasma concentration and the content of ubichinon in the muscles.

The general effects of ubichinon include increasing muscle performance, stabilising the immune system, raising cellular energy utilisation and removing free radicals. In hospitals this substance is used therapeutically to treat infarcts.

The use of **ubichinon** in sport is controversial. It has not been possible to document an increase in performance through ingestion among athletes (BRAUN et al., 1991; WESTON et al. 1997). What is certain is that consumption of 1 mg/kg of ubichinon over 28 days leads to a significant rise in blood concentration from 0.91 to 1.97 μg/ml (WESTON et al., 1997). It is possible that through ingestion of ubichinon a general ability to handle load as well as cardiopulmonary endurance in sport practitioners, is increased. Further studies among athletes are needed on the effects of the vitamin-like substance ubichinon or coenzyme Q 10 , especially with regard to its function as a neutraliser of free radicals.

8.4 Caffeine

Caffeine is the active ingredient in coffee, tea and guarana. As a luxury article, caffeine is used by many people regularly for mental and physical stimulation. Caffeine activates the central nervous system and the sympathetic nervous system. Through the activation of the catecholamines by caffeine, increased numbers of Free Fatty Acids are released and their utilisation increased. This simultaneously impedes glycogen breakdown. Scientifically the performance increasing effects of caffeine are only based on the ingestion of pure caffeine. Studies with coffee are not representative. The effect of caffeine depends on the dosage.

Caffeine	
Physiology:	• Activation of CNS and sympathetic nervous system • Increases effect of the catecholamines • Increases release of Free Fatty Acids (FFA) and increases their oxidation • Impeding of glycogenolysis (lowers carbohydrate utilisation)
Therapy:	• 1 to 6 mg/kg (100-300 mg) • Over 15 mg/kg side-effects (nervousness, sleeplessness, tachycardia, increased blood pressure, gastro-intestinal problems)
Sport:	• Improves reaction time and coordination (200-300 mg) • Improves longer endurance performances (250-350 mg; 5-6 mg/kg) • Does not influence strength endurance and short endurance performances
Doping:	• Limit value positive: 12 μg/ml urine • In 2-3 hours 200 mg caffeine lead to a urine concentration of 3 μg/ml • Practical limit doses are 300-400 mg/day

Table 45

Ingestion of 300 mg of caffeine **improves reaction time** and **biomotor coordination**. From ingestion of 6 mg/kg (about 350-400 mg) upwards a positive influence on **endurance performance** can be expected. Influencing strength and short endurance performance requires the greatest ingestion of caffeine; here quantities of over 7 mg/kg are assumed. After ingestion of 9 mg/kg of caffeine an hour before loading a significant increase in cycling and running performance in well trained athletes was documented (GRAHAM/SPRIET, 1991). With this quantity of caffeine the **doping limit** was narrowly missed.

Because caffeine always has a performance improving effect in sport it was determined that a quantity of 12 μg/ml of caffeine in urine constitutes a doping offence. This border amount can be reached by drinking more than four cups of strong coffee. A normal coffee contains 300 mg/l (or 120 mg per large cup) and a strong coffee or espresso contains 500-600 mg/l of caffeine. There are obviously great individual variations in the digestibility and breakdown of caffeine.

The caffeine doping limit is reached with certainty when more than about 9 mg/kg of body weight of caffeine or products containing caffeine are ingested. This means more than five cups of coffee for a person weighing 70 kg.

Care should be taken with the consumption of brain drinks (disco drinks) containing carbohydrates and caffeine which are available on the market as "Red Bull" or "Flying Horse". The products made in Germany (250 ml cans) contain 320 mg of caffeine and thus the quantity found in about three cups of coffee. Austrian products contain considerably more caffeine.

If 2-3 cups of normal coffee are drunk two hours before the start there is no problem with regard to exceeding the urine borderline value.

8.5 Alkaline Salts

There has been an interest in buffer substances in sport for over 60 years (DENNING, 1937). High lactate concentrations occurring as a result of intensive exercise were the reason for reducing these by consuming physiological buffer substances. The natural buffer capacity consists of carbon dioxide bicarbonate buffer (53%), haemoglobin (35%), phosphate and serum proteins. The still unsolved problem with the ingestion of acid binding substances is their poor digestibility in the intestines. These include sodium bicarbonate and sodium citrate. These need to be ingested in great quantities in order to have a buffer effect. The bicarbonate buffer is mostly used when acid develops in the muscle. Drinking bicarbonated mineral water has a positive influence on the binding of hydrogen ions and can thus positively influence sporting performance capacity.

Another buffer system that is of interest is phosphate. Phosphorus is an indispensable element in nutrients and works biologically in the form of its salts **(sodium phosphate, potassium phosphate)**. Phosphorus is a major component in the central substances of energy transformation, ATP and creatine phosphate, the oxygen dissolving effector in the erythrocytes (2,3-diphosphoglycerate or 2,3-DPG) and in various B group vitamins. It therefore seems reasonable to assume that through additional ingestion of phosphate salts (e.g. 4 g of sodium phosphate/day) there could be a positive influence on performance capacity. The evidence of the performance increasing effects of phosphate salts is contradictory (PARRY-BILLINGS/M.MACLAREN, 1986; MACLAREN/ADAMSON, 1995). There is also the fact that the training-induced increase in maximum oxygen intake also increases buffer capacity (OSHIMA et al., 1997). Ingestion of alkalising salts to influence performance is normally not usual in sporting practice. Recently there has been evidence that as a result of ingestion of a potassium-iron-phosphate-citrate complex, through ammonia buffering the oxidational ability of the liver for lactate is raised. This ammonia buffer could attain practical significance for intensive exercise which is coupled with a rapid rise in lactate.

8.6 Creatine

Creatine is a physiological active substance which is indispensable for muscle contraction. The **body's own production** of creatine occurs from the three amino acids arginine, glycine and methionine in the liver, kidneys and pancreas. Additionally about 1 g is ingested daily with food. The **main sources** are meat and fish. In a man weighing 70 kg, daily utilisation of creatine is about 2 g (BALSOM et al. 1994). In order to ingest 5 g of creatine, 1.1 kg of raw beef would need to be eaten. Vegetarians ingest practically no creatine through vegetable foods and are thus entirely dependent on self-synthesis.

The **plasma concentration** of creatine is between 20-100 μmol/l. In vegetarians it is low at 25.1-32.4 μmol/l (DELANGHE et al. 1989). The creatine muscle depots of vegetarians, however, are not clearly smaller (HARRIS et al. 1992). Animals given creatine analogous feed (beta guanidine propionic acid) display a drop in muscle creatine of 65% (ADAMS et al. 1994). From this it can be concluded that the greater proportion of creatine is ingested through food. The creatine pool is 120-140 g of which 95% is stored in the muscles. Only 30% of muscle creatine is free, most of it is incorporated in the alactic energy store creatine phosphate (CP). In the M. vastus lateralis the average **total creatine** concentration at rest is 118.1 ± 3.0 mmol/kg TG (HARRIS et al. 1992). The fast twitch muscle fibres (FTF) have a higher total creatine content than the slow twitch fibres (STF). In untrained men average creatine phosphate was 75.5 ± 7.6 mmol/l TG and creatine was 48.0 ± 7.6 mmol/l TG (HARRIS et al. 1992). The creatine used up in muscle contraction is replenished via the bloodstream. Muscle acidity favours the transformation of creatine into the anhydrid, **creatinine**. Creatinine increases in the blood during intensive interval loads and is excreted. Produced mainly in the liver, creatine gets into the muscle cells via an active membrane transport. The high **creatine requirements for the production of creatine phosphate** for speed strength performances and intensive short loads leads to a decrease in serum creatine, as was documented in interval loads of 100 m to 1000 m (SCHUSTER et al. 1979). Sprinters have a higher ATP and total creatine content in the muscles than long distance runners. Analogous to this, track cyclists who load themselves more intensively have a higher creatine content than road cyclists as earlier studies by the author have shown. This documents the ability of the creatine stores to be trained using short duration alactic training.

There is a constant exchange of creatine between the cytosole and the mitochondrium which is needed for production of CP in the mitochondrium and resynthesis of ADP (adenosine diphosphate) to ATP (adenosine triphosphate). The

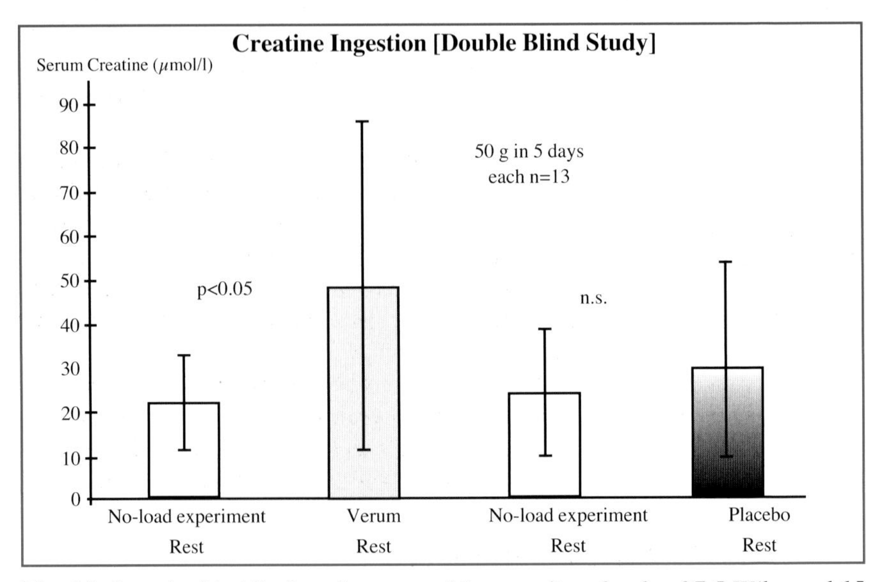

Fig. 28: In a double blind study repeated intermediate loads of 7.5 W/kg and 15 seconds duration within an endurance load led to a significant increase in alactic performance capacity after ingestion of 50 g of Creatinabol(r) in five days (5 x 10 g). Own unpublished findings.

mechanism of cytosolic and mitochondrial ATP resynthesis and the stabilisation of the CP pool in the cytostole is made possible by a special exchange mechanism called the creatine phosphate shuttle (BESSMAN/ SAVABI, 1988). The **creatinine shuttle** improves provision of ATP and CP and at the same time restricts the metabolisation of creatine to **creatinine**. The creatinine leaves the muscle cell and is excreted renally. Because the mitochondrial CK isoenzyme rephosphorises the muscle creatine back into CP, the creatine requirements from the body's own production and the supply through food are reduced. Through the rise in creatine concentration in the blood after increased ingestion, a reverse pressure is applied to the muscle cells and these absorb more creatine. Creatine phosphate can buffer the increased numbers of hydrogen ions occurring during intensive loading which arise through lactate production. This extends the fatigue limits of the muscle and creates the prerequisites for total anaerobic performance capacity to increase (MAUGHAN, 1995)

Ingestion of higher doses of creatine over several days influences short duration alactic performances. In trained athletes an **increase in alactic performance capacity** is probable **with high creatine supplementation**. The prerequisite is a high **dosage of creatine of 20 g/day over five days** to saturate the store (HARRIS et al., 1992; BALSOM et al. 1993; GREENHAFF et al. 1993, 1996; HULTMAN et al. 1996). According to the author's own research, **low saturation dosages** of creatine, e.g. 6 to 10 g/day over five days, also have a positive effect on performance capacity (Fig. 28).

GREENHAFF et al. 1994 and CASEY et al. 1996 found in about 30% of athletes there was not a sufficient increase of total muscle creatine after supplementation of 5 x 20 g (100 g). An increase of the phosphocreatine stores of over 8% is seen as the minimum prerequisite for an increase in alactic performance capacity. This is evidence that some athletes do not react to creatine ingestion, the so-called **non-responders**. Creatine ingestion only leads to performance improvement in those athletes who can actually produce more creatine phosphate.

There is a close relationship between creatine ingestion and carbohydrate metabolism, for simultaneous ingestion of creatine with 370 g of carbohydrates/day increases the depositing of creatine in the muscles (GREEN et al. 1996). Effective dietetic creatine preparations take these physiological findings into consideration.

8.7 Other Biological Substances

In addition to the substances briefly described here there are others which have an influence on performance capacity, regeneration or the state of well-being. Similar to the effects of some medicines, these performance aiding substances, also called **ergogenic**

substances, influence the psyche. Possibly their effects unfold under this aspect. With the following substances research following scientific criteria still needs to be done. This does not mean they have no effects. The ergogenic substances described here are not on the list of banned substances and are thus not doping (Table 46).

Aspartates

Aspartates are salts of the amino acid asparagine, and in the form of potassium or magnesium aspartate have an effect on the increased breakdown of ammonia. By lowering ammonia concentration fatigue during load can be influenced.

Inosine

Inosine is a nucleotide with a metabolic function in the production of ATP and the splitting of muscle glycogen. Performance increasing effects have not been documented.

Pollen

In pollens there are mixtures of vitamins, minerals, amino acids and other biological active substances, although in very small quantities. Pollens are the male germ cells of plants and are also considered the original substance of life. Seafarers in former times considered honeycombs with pollen to be a tonic. Competitive athletes use pollen preparations to aid regeneration.

Gelee Royal

The juice used to feed the queen bee is a jelly substance produced by the worker bees (Gelee Royal). The ingredients have a great effect on the rapid growth of the queen. The hormone-like ingredients probably influence regeneration and are used outside sport for revitalisation.

Ginseng

The steroid-like ingredients of the Asian ginseng root stimulate the functions of the central nervous system, activate the biomotor system and are favourable to concentration. These properties are useful for some sports such as combat or game sports. When dosed extracts are ingested, general readiness to perform is influenced positively (see Table 46). Overdoses lead to sleeplessnes, high blood pressure or diarrhoea. It is possible that regeneration is aided by ginseng extracts, as ginseng also has antioxidative effects.

Wheat Germ Oil

Wheat germ oil was already being taken over 1,000 years ago to maintain health and performance capacity. The effects were originally attributed to the plentiful amounts of vitamin E in the germ oils (see chapter 6). The long chain alcohol octacosanol, however, has been identified as the active substance in wheat germ oil. American scientists attribute to octacosanol an improvement in the supply of energy and thus an influence on endurance performance capacity. Scientific checks have not been able to confirm this effect.

Yeast Cell Preparations

The healing effects of yeasts has been known for a long time (PIENDL/RUMMEL-PITLIK, 1985). Enzymatically split yeasts contain a range of valuable ingredients. These include the vitamins of the B complex, trace elements (selenium), minerals (magnesium, calcium), ubichinon, glutathione, complex carbohydrates, enzymes among others. Because these substances influence the cell components desoxyribonucleic acid (DNA) and cell ageing they have become scientifically interesting again. In this composition yeasts have an immunising effect and influence metabolic processes such as the antioxidative cell potential. Through ingestion of an enzyme yeast preparation over six weeks it was possible to describe effects in the direction of better stress management and regeneration (BERG et al. 1997).

Performance Enhancing with Ergogenic Aids *	
• **Carbohydrates (CH)**	40-80 g/h CH during competition (C). Before C 3-6 g/kg CH.
• **Caffeine**	200-500 mg before C. At 9 mg/kg the doping limit is reached. (Drinking coffee contains 50-90 mg of caffeine in 100 ml)
• **Creatine (CR)**	Altogether 30-100g in five days is enough to fill up the creatine stores of the muscles by 20%. The maintaining dose for alactic training is 2 g/day for a limited time. In about 20% of athletes CR does not have any effect (Non-responders).
• **Alkaline Salts**	Sodium bicarbonate (citrate) 0.3-0.5 g/kg in a litre of water buffers hydrogen ions (lactate). In this form buffer substances are usually indigestible. Better: Potassium-iron-phosphate-citrate complex to buffer lactate.

Tab. 46 Scientific experiments prove that there is an increase in performance.

9 Prohibited Substances and Methods (Doping)

People training, like other citizens, have the right to take medical drugs when they get sick and to get medical treatment. By additional ingestion of certain substances the parameter for securing load tolerance can be extended. Sporting load is often restricted by the taking of medicines. Ingestion of medicines or other active ingredients is not doping as long as they are not on the doping list. Taking of medicines when there are comprehensible indications reduces sporting load tolerance and should be closely examined in the case of competition obligations (ENGELHARDT/NEUMANN, 1994). One must, however, always differentiate between illness and the effects of medical drugs on sporting performance capacity. If there are doubts, a start on medication should be refrained from. Performance capacity has been shown to be diminished by frequent taking of antibiotics for the treatment of infections.

The **term doping** commonly used in competitive sport is currently being worn thin (Table 47). In the Federal Republic of Germany, in the German Sports Federation (Deutscher Sportbund, DSB) the guidelines of the Medical Commission of the International Olympic Committee (IOC), which classifies substances considered as doping in competitive sport, have applied since 1977. The basis for **sanctions** against athletes is the **doping list**. For **squad athletes**, a **refusal** to undergo a urine test is immediately considered doping.

The **definition** of doping according to the IOC is as follows:

1. Doping is the attempted unphysiological increasing of an athlete's performance capacity through use (ingestion, injection or giving) of a doping substance by an athlete or a helping person (e.g. team captain, coach, advisor, doctor, nurses or masseurs) before a competition and anabolic and peptide hormones outside competitions too.

2. Doping substances in the sense of these guidelines are in particular stimulants and related substances (amphetamines, ephedrine, metaprotenerol), narcotics (morphine, diamorphine), anabolic agents and peptide hormones as well as masking substances (e.g. diuretics). Sport-specifically other substances, e.g. alcohol, sedatives, psychiatric drugs, can be listed as doping substances (Table 48).

Doping in Sport

1. In 1964 the IOC decided to ban the taking of medicinal drugs during the Olympic Games.

2. The list of prohibited substances which make it possible to improve performance is constantly updated by the Medical Commission of the IOC.

3. "Doping is the use of substances from the prohibited substance groups and the use of banned methods.
   The doping list can be changed at short notice if there is a reason to do so."

4. Proven taking of prohibited substances, manipulation or refusal to provide urine samples is punishable.

Table 47

The DSB has set up a **doping list** which applies for the **sports associations**. The doping list is constantly updated. Thus e.g. in 1998 **androstendione**, a preliminary stage of testosterone, was put on the banned list. In addition to the anabolic effects, androstendione causes masculinisation of women, hair loss and a deep voice (ABRAMOVICZ, 1996). In 1998 the stimulant **carphedenone** was also placed on the list of banned substances, as was bromantane in 1997.

In addition to doping controls in competitions, now **training controls** at short notice are increasingly being carried out in Germany. Non-permitted taking of anabolic steroids is especially pursued.
According to DONIKE/RAUTH (1992) the following **substance groups and methods** are on the banned list:

1. Prohibited Substance Groups
A. Stimulants
B. Narcotics
C. Anabolic steroids
D. Beta blockers
E. Diuretics
F. Peptide hormones and analogues.

2. Prohibited Methods
A. Blood doping
B. Pharmacological, chemical and physical manipulation

3. Substance Groups, only permitted with certain restrictions
A. Alcohol
B. Marijuana
C. Local anesthetics
D. Corticosteroids
E. Specified Beta-2 Agonists

If an athlete is undergoing **medical treatment** and taking medication which is on the doping list , he is normally **not fit for competition**. As not all doctors are aware of the prohibited list, if an athlete is taking a substance unknown to him he should make the doctor aware of the requirements of the doping list. Under the aspect of medical indications there is currently no necessity for the taking of stimulants (amphetamines) or anabolics by competitive athletes, even in cases of illness.

Information on active substance ingestion: Anti-Doping Commission DSB and NOK, Otto-Fleck-Schneise 12, D-60528 Frankfurt/M., Germany

In the doping control after competitions or in training, squad athletes are required to put down in writing the medication they are taking (e.g. cough medicines, sleeping drugs, antibiotics etc.)

Analyses of urine samples (A and B samples) are only possible in the special laboratories licensed for this purpose by the **IOC**. In Germany only the **doping laboratories Cologne** and **Kreischa** (Dresden) have the authorisation to carry out analyses.

As solid and lasting improvements to performance are only possible through training anyway, for ethical and moral reasons, out of fairness to fellow sporting competitors, and because of the health risks, as a matter of principal taking active substances and medical drugs which are on the prohibited list should be shunned.

Even with the increasing commercialism and professionalism, as well as the pressure to succeed on both the national and international level in competitive sport, it is possible to achieve sporting performance through the **use of natural and physiological performance** means. The decisive means lie in sport-scientifically planned and supported **training**, in the optimisation of **regeneration**, in maintaining a **diet** and **supplementation** suited to sport, and in the use of **altitude training and climatic training**.

Prohibited Substances and Manipulations in Competitive Sport*

Effect Group	Effects
1. Stimulants (amphetamines, ephedrines, cocaine, fencamfamine, fenetylline, pemoline). Since 1998 carphedenone	Major activation of the central nervous system, removal of fatigue; shifting of performance limits
2. Narcotics (morphine, pentazocine, pethidine)	Euphoria, suppression of pain, especially in the muscles; further loading free of pain
3. Anabolic Substances (clostebol, mesterolone, nandrolone, stanozolol, testosterone, clenbuterol). Since 1998 androstendione	Muscle development, fibre hypertrophy, euphoria, mascularisation of women, impotence in men; increased strength, great training enjoyment
4. Diuretics (furosemide, acetazolamide, spironolactone, triamterene)	Major urine excretion, urine dilution; dehydration in weight class sports; start in lower weight class in competition
5. Peptide Hormones and Analogues [growth hormone (GRH), erythropoietin (EPO), corticotrophin (ACTH)]	Muscle growth and whole body anabolism through GRH and ACTH, increase in erythrocytes (haemoglobin increase) through EPO; increased strength, faster anabolism, increased oxygen transport capacity (EPO)
6. Analogue Active Substances (Chemically and pharmacologically related compounds of 1-5 above)	Similar effect to the compounds listed under 1-5.
7. Banned Methods (Pharmacological, chemical and physical manipulation of urine tests, blood doping)	Disguising or greatly diluting active substance consumption; increased oxygen transport capacity through increase in haemoglobin
8. Substance Groups with Restrictions (Alcohol, beta blockers, local anaesthetics, asthma drugs, corticosteroids, marijuana)	Ingestion or application notifiable. Banned in certain sports (e.g. alcohol and beta blockers for shooters)

*Table 48: * Refusal of doping tests in training or after competitions is considered a positive finding.*

10 Outsider Diets

The wide range of options for reducing body weight or body mass is not comprehensible for either doctors, patients or normal citizens. Table 49 contains an overview of selected **reduction diets**.

Modern living gradually leads to excess weight in men, women, adolescents and children if they are physically inactive. If the inactive lifestyle is maintained, over ten years just consuming 2% more energy (calories) per day leads to being 10 kg overweight. Being overweight is tolerable to a certain degree.

When men exceed 23% and women exceed 30% of the normal weight in relation to their physical build, their health risks rise considerably. These figures are almost identical with the **Body Mass Index (BMI)** of over 23 for women and over 25 for men (see chapter 12). With a BMI over 30 any further weight increase must be avoided. The probability of premature illnesses in the cardiovascular system and metabolism increases considerably with this excess weight and obesity (adipositas). The life expectancy of obese people is usually shortened on average by about five years. On the other hand, if leanness is too great, i.e. the BMI is under 20 in adults, there is also increased proneness to illness.

In addition to the health significance, these known facts of the side-effects of being overweight are the starting point for the propagation of possibilities for reducing weight.

Reducing body weight can be achieved in principle with the following **dietary measures**:

1. Total fasting
2. Energy reduced mixed diet
3. Diets with extreme nutrient ratios
4. Industrially produced nutrient mixtures (formula diets)
5. Modified fasting
6. Fruit or rice days
7. Outsider diets, such as Hollywood diet, wholemeal diet, Hay food combining, Atkins diet, potato diet, Cambridge diet etc.

Thirst under control

POLAR heart rate monitor, Fitwatch sender and receiver

Bananas again and again!

German triathlon champion 1995 Ralf Eggert analysing training load

Written invitation to the doping contol

Most diets are geared to an **energy intake of about 1,000 kcal/day**. In this way **basic utilisation** of 1,500-1,700 kcal/day falls short by about 40% and an **energy deficit** is produced. With these diets a reduction of body mass of 1 kg per week is certainly possible. The initial successes in weight reduction results from water loss, and are not typical for stable and safe reduction of body mass.

Numerous studies lead to the conclusion that **dietary measures without a simultaneous increase in physical activity are not very effective**.

In the initial phases of the dietary measures, two hours of sport/week is enough. Exercise should be chosen in such a way that the heart rate is 130-140/min. At this moderate exercise level about 400 kcal/h are additionally used. An example of possible weight reduction using a formula diet and sport with various patient groups can be seen in Fig. 29.

If one succeeds in keeping to a daily energy deficit of 600 kcal and total energy intake is 1,200-1,800 kcal/day, then it is possible to stably lose 0.5 kg after three months, and 1 kg after six months without sport. In any case one should ensure that the resolution to lose weight does not become the beginning of psychologically based **eating disorders**. Under clinical conditions, in cases of adipositas (obesity) formula diets with 400 kcal/day are undertaken successfully.

The danger with extreme diets is that deficiencies of minerals, vitamins and other active substances occur (see Tables 35 and 38). When choosing the options for weight reduction, attention should always be paid to this. Reducing energy intake to 1,000 to 1,500 kcal/day combined with regular sporting load of about two hours per week has the most certain effect (Tables 50 and 51). Staying at a rehabilitation centre helps. Under guidance, the change of surroundings and optimum conditions for practising sport e.g. swimming pool, cycling tracks, running terrain, can ease the start of regular sport. Experience with the application of extreme methods for weight reduction shows that the former weight soon returns if the lifestyle is not changed fundamentally. Weight reduction is successful if the reduction of 5% of original weight is maintained for over a year. Experts dealing with the reduction of body weight agree that the forms of short term and **drastic weight reduction cannot be the objective**. The main problem after every reduction of body mass is maintaining the lower body weight achieved.

Diet Type	Principle	kcal/day	Weight Reduction	Advantages	Disadvantages
Total fasting	Zero diet 3 l water, vitamins and minerals	0	1st wk 5-6 kg 730-810 g/d 2nd wk 2.5-3 kg 400 g/d	Major weight reduction	Inability to work, in-patient, protein catabolism of 25 g/d
Modified fasting	One-sided protein ingestion; (curd cheese, egg white) + 3 l fluids	300-400	1st wk 4-5 kg 2nd wk 2 kg	Major weight reduction, little protein catabolism	One-sided nutrient ingestion, impeded ability to work, mostly as in-patient
Energy reduced mixed diet	Reduced energy intake Qualitatively full valency diet 70 g proteins (28%) 40 g fat (37%) 85 g CH (35%)	1000-1500	Wk 1 kg	Sensible diet over longer periods, improved eating habits, individual variable food plans	Keeping records, weighing and measuring food, financial requirements for programme and membership of "Weight Watchers"
Variation in practice: "Weight Watcher"	15% fat, 45% CH, protein, small servings Minerals and vitamins	Stages I-III 1040-2090	Wk 1 kg		
Formula diet "CAMBRIDGE"	Industrially produced nutrient mixture with defined mixed composition, completely balanced, 2 l fluids	750 (3 x 250) 1000-1500 diet + 1-2x250	1st wk 1.5-2 kg 2nd wk 1 kg	Ready to eat, easy preparation, balanced food, vitamin and mineral ingestion	Few flavour variants, liquid, no major change in eating habits
Combination diet	Modify fasting and formula diet Energy reduced mixed diet and formula diet				

Diet Type	Principle	kcal/day	Weight Reduction	Advantages	Disadvantages
Diets with extreme nutrient ratios "Outsider diets"	Elimination of a main nutrient				
e.g. "POINT DIET"	High fat and CH low (points)	max. 60 points	1st wk. 2,5-3 kg	Rapid weight loss through CH ban, strong lipolysis	Points and Atkin diet: negative mineral and vitamin balance ketotic metabolism, cholesterol increase, rapid body mass after ending
"ATKIN DIET"	1st wk. no CH 2nd wk. 5 g CH/d (vegetables, fruit, wine!)	As much fat and protein as desired			
"POTATO DIET"	All meals with potatoes and vegetables	1000	1-2 kg/wk	Always feel full, rich in potassium and fibre, good taste	Low offering of minerals, vitamin and unsaturated fatty acids. Deficiency : Vit. B, Ca, Fe, Zn and Mg
"HOLLYWOOD DIET"	Tropical fruits	800-1000	1-2 kg/wk		
Short term diets					
"FRUIT DAYS"	Daily 2 kg fruit and drinks in 5-6 servings over 1-2 days	1200	Dehydration and 200 g body mass reduction/day	Purification, pectine as fibre like ballast substances	Nutrient deficiency if used continuously
"RICE DAYS"	200 g rice and 500 g fruit over 1-2 days and drinks	950	Dehydration and 200 g body mass reduction/day	Strong diuresis through potassium rich food	Not a long-term diet
Energy reduced foods	Industrial products: low fat, , low CH fibre rich (powder, granulate ,muesli, flakes, sweetener)	1000-2000	0.5-1 kg/wk	Full feeling, low calorie foods	Monotonous taste

Table 49: Overview of diets which reduce body mass (NEUMANN, 1991)

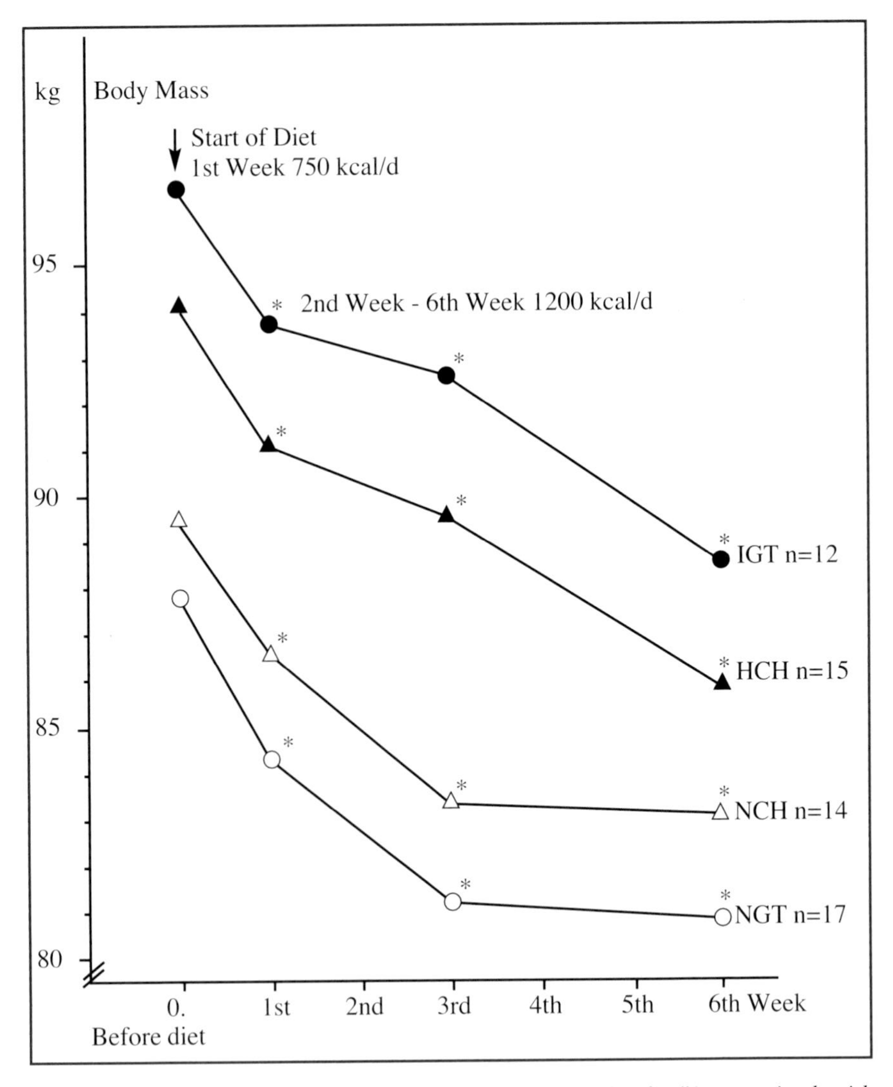

Fig. 29: Behaviour of body mass when the "Formula diet" is practised with patients with varying risk factors. IGT = Interrupted Glucose Tolerance (preliminary stage of Diabetes mellitus). HCH = High Cholesterol. NCH = Normal Cholesterol. NGT = Normal Glucose Tolerance (normal insulin regulation). With the prescribed diet patients with the risk factors IGT and HCH and the greatest weight lose the most weight

Measuring Parameters	Starting Value	After 3 weeks	After 6 weeks
Body mass (kg)	82,2±9,8	78,5±8,9**	77,0±8,9**
Body Mass Index (kg/cm²)	29,4±2,4	28,5±2,3**	28,0±2,4**
Fat (kg)	25,8±5,1	23,6±4,6**	22,4±4,7**

** $p < 0.001$ of starting value

Table 50: Changes in anthropometric measurements after food restriction (1,000-1,500 kcal//d) and two hours of sport/week over six weeks with 15 overweight women (1.66 ± 0.07 cm body height)

Measuring Parameters	Starting Value	After 3 weeks	After 6 weeks
Glucose (mmol/l)	5,4±0,7	4,5±0,8*	4,3±1,2*
Free Fatty Acids (μmol/l)	427±149	765±227*	645±221*
Cholesterol (mmol/l)	4,70±0,8	4,95±0,6**	5,11±0,6**
Serum urea (mmol/l)	5,5±1,2	4,2±0,7*	4,4±0,9*
Uric acid (μmol/l)	319±44	275±81*	275±45*

* $p < 0.05$ of starting value

Table 51: Changes to metabolism measuring parameter after food restriction (1,000-1,500 kcal//d) and two hours of sport/week over six weeks with 15 overweight women (see Table 50)

11 Other Forms of Nutrition

11.1 Vegetarian Nutrition and Sport

In the course of development, human dietary habits have changed many times. Humans were **originally vegetarians**. Then came transitions to consumption of animal foods. The vast majority of the earth's population is now dependent on **mixed food**. The ability to eat a one-sided diet, however, still remains. Whereas the predominantly meat eaters, e.g. Eskimos or shepherds in certain regions, are not so much in the focus of dietary forms, this does not apply to vegetarians.

For competitive athletes the discussion about the pros and cons of a vegetarian diet is of practical significance. If there is an increase in sporting load of more than ten hours/week, a vegetarian diet leads to certain nutritional deficiencies. Initially **vegetarianism** was seen as an alternative dietary form with performance enhancing potential, and the number of athletes who tried to eat a vegetarian diet increased temporarily. The great performances of vegetarian athletes in the "Deutschlandlauf" in Germany in 1987 became widely known. In this run, 1,000 km were covered in 20 days in average stages of 50 km/day. The dietary study carried out on this occasion (EISINGER, 1990) showed no performance differences between runners eating ovo-lacto-vegetarian whole food and those on a normal mixed diet. In the final analysis both the conventional diet runners and the ovo-lacto-vegetarian whole food athletes survived the high load. From a nutritional physiology point of view there is no reason to criticise athletes who eat a different diet, and can consciously compensate deficits with supplementation.

It should be noted that there are **various gradings** in **vegetarianism**. The strictest form of vegetarianism is practised by the vegans who avoid absolutely foods of animal origin. The **lacto-vegetable diet** is encountered more frequently. These vegetarians additionally consume milk to compensate the protein and mineral deficit. If in a vegetarian diet eggs are consumed as well as milk, this dietary form is called an **ovo-lacto-vegetable diet**. A characteristic of strictly vegetarian nutrition is abstinence from alcohol and nicotine.

As a variation on vegetarianism **whole food nutrition** has been developed (WORM/SCHRÖDER, 1987; Wolfram, 1988). According to the theory of whole food nutrition, five stages are differentiated:

1st Stage: **Unaltered foods** e.g. grain sprouts, fresh vegetables, fruit, oils, seeds, nuts, milk.

2nd Stage: **Dressed foods** e.g. grains, chopped up vegetables, sauerkraut, curd cheese, mineral water, herb tea, dried fruit.

3rd Stage: **Temperature treated foods** e.g. baked goods and soups made from wholemeal flours, heated vegetables, stewed fruit, oils, milk and dairy products, mineral water, barley malt coffee substitute, cocoa.

4th Stage: **Processed foods** e.g. baked goods from ground grain, canned fruit and vegetables, margarine, oils, milk and dairy products, meat, fish, eggs, tap water, ground coffee, tea, beer, wine.

5th Stage: **Finished products** e.g. grain, sugar-energy bars, vegetables, fruit, oils, fats, nuts, milk and dairy products, vitamins, alcohol, isolated food substances, amino acid preparations, meat, fish, eggs, all beverages, sweeteners.

The **principle of whole food nutrition** is geared to consumption of nutrients of stages 1 to 3 with the idea that **50% fresh raw foods** and **50% heated foods** should be consumed. Biological growing of foods which enter the market as bio products is linked with the practising of whole food nutrition. This biological growing of grain, vegetables and fruit, however, does not offer a complete guarantee against harmful substances. Comparisons of normally produced foods with those produced biologically showed no differences. The undesired accompanying substances e.g. heavy metals were distributed practically evenly.

Both conventional mixed food and vegetarian food can lead to vitamin and mineral deficiencies, but these vary. In chapters 6 and 7 these problems are discussed.

Analyses of the deficits of athletes on vegetarian or traditional diets are very much characterised by coincidences in the selection of the persons investigated and are accordingly contradictory.

In competitive sport there are a number of successful athletes who are vegetarians. They eat predominantly ovo-lacto-vegetarian food. They consciously compensate possible deficits of active substances and vitamins, caused by practising their sport, by supplementing. Athletes eating whole foods usually do better than traditional eaters in the use of nutritional advantages for the simple reason that they pay more intensive attention to their diet. This applies to e.g. comparisons of the iron, magnesium or zinc balance; here vegetarians are often better supplied than normal eaters. The calls for an increase in the proportion of carbohydrates in nutrition found in many specialised nutrition books are best met by vegetarians (JUNG, 1984; HAAS, 1986 and 1986; HAMM/WEBER, 1988; KEUL/WITZIGMANN, 1988; PETERSON/PETERSON, 1988; WORM, 1988; GEISS/HAMM, 1990; EISENMAN et al. 1990; CLARKS, 1990; HAMM, 1991; KONOPKA, 1994 among others).

One-sided dietary orientation to carbohydrates in competitive sport training is no longer absolutely valid because of new findings with regard to preventive health. The saturated and unsaturated fatty acids are often underestimated with regard to their significance in metabolism and thus for competitive athletes. For sport practitioners in competitive sport alcohol should not be a topic; in practice the situation is more sobering. For hobby athletes, properly managed alcohol consumption, especially in the form of red wine (1-2 glasses/day) has a protective effect on the blood vessels. According to new findings on preventive health, modest alcohol consumption is supposed to prevent hardening of the arteries. The components of red wine have a proven antioxidative effect.

Vegetarians training in competitive sport only represent the mildest form of vegetarianism because they also consume animal dairy products and eggs. The **vegans** totally avoid animal products, they only eat foods of plant origin, without milk, dairy products and eggs and are rarely active in competitive sport. Apart from having deficiencies of iron, vitamin B_{12} and biologically high valency proteins, objectively they are only loadable within limits. Their muscular strength potential can hardly develop in training without additional ingestion of proteins, concentrates of which are mainly made from milk, whey or fish. A possible alternative might be to take a protein concentrate made from soy flour.

Whole food nutrition, which is based on limited industrial handling of the products (milk, grain, vegetables among others) is definitely compatible with competitive sport training. By regularly **checking the mineral balance**, or the vitamin status, deficits can be discovered and compensated. Plant (soy) and milk based protein concentrates help compensate a possible deficit of biologically high valency amino acids (see Table 13). The problem thus lies not in individually structured dietary habits but rather in knowledge about limitations and potential bottlenecks of this or that dietary form and the energy requirements of the sport practised. The increased energy needs of a top athlete of 6,000 to 8,000 kcal/day not only represent calorie demands but also quality demands on food, such as **high nutrient density**. By this the mineral, vitamin and fibre content in a particular quantity of food is meant. In addition to energetically securing training load, sports nutrition is of great importance for accelerated regeneration and has a major influence on performance capacity. Current knowledge indicates that nutrition of athletes in competitive sport can no longer be left up to subjective feeling. Successful athletes increasingly take into consideration scientific findings on sports nutrition and use these for themselves.

11.2 Nutrition and Sport in Cases of Diabetes mellitus

In healthy people eating food leads to an increase in blood sugar. The increase in blood glucose concentration signals to the pancreas to release insulin. At rest glucose ingestion and corresponding insulin production are in well adjusted regulation. The higher the blood sugar rises, the more insulin the pancreas produces.

The **sugar related illness** or **Diabetes mellitus** occurs when the blood sugar level no longer, or only slowly, goes down. There are two forms of diabetes. One form is the **insulin-dependent diabetic** (type Diabetes I). This type I diabetic must regularly inject insulin when the illness occurs and is constantly medically checked. This illness occurs early and can affect children and adolescents. Numerous examples show that **insulin-dependent athletes** who maintain their treatment, and the corresponding training with the appropriate discipline, can also reach the field of top sport.

Of the almost six million diabetics now in Germany, type I diabetes makes up about 10%. 90% of diabetics are type II. **Type II diabetes is not insulin-dependent**. Here there is usually an insulin sensitivity in the tissue and the low **insulin secretion** of the pancreas does not meet the demands of high carbohydrate consumption. **Insulin production does not meet requirements**.

The criterion for a diabetes screening is the blood glucose concentration.

Normal blood glucose:	<5.3 mmol/l (<95 mg/dl)
Borderline blood glucose:	5.3-6.1 mmol/l (95-110 mg/dl)
Pathological blood glucose (rest):	>6.1 (>110 mg/dl)

If two hours after being given 75 g of carbohydrates the blood glucose rises above 11.1 mmol/l (> 200 mg/dl), there is a case of Diabetes mellitus or a major carbohydrate tolerance disturbance. This **carbohydrate load test** is carried out to confirm the diagnosis when there are borderline findings.

Here only normal weight and partially **sporting type II a diabetics** are discussed, who make up about 20% of the type II diabetics. For **type II diabetics endurance training is almost an ideal treatment**. Through the physical training the blood glucose gets into the tissue via the insulin-independent receptors and the raised blood glucose level is lowered without insulin or blood sugar lowering medication.

Fig. 30 shows an example of how a young endurance athlete suddenly became a type II diabetic as a result of a viral infection. After an endurance load of 240 W over 105 min his blood sugar level decreased considerably. The athlete then drank a highly dosed complex carbohydrate. His release of insulin was too low to deal with the amount of glucose in his blood. The result was that after 90 min his blood glucose rose to 18 mmol/l (324 mg/dl) and slowly decreased again. This state was not threatening for the athlete, he felt good. The practical consequence of this example is that sporting diabetics should not consume large quantities of carbohydrates at once after exersice. It is better to distribute ingestion of carbohydrates over quantities of 24 to 48 g (2-4 BU). The formerly common Bread Unit (BU) corresponds to ingestion of **12 g of digestible carbohydrates**. The BU is now called the **CHU or CU (Carbohydrate Unit)** and **corresponds to 10 g of digestible carbohydrates**. Even during exercise, glucose quantities of 3-5 CU are sufficient as glucose absorption in the intestines is only 60 g/hour.

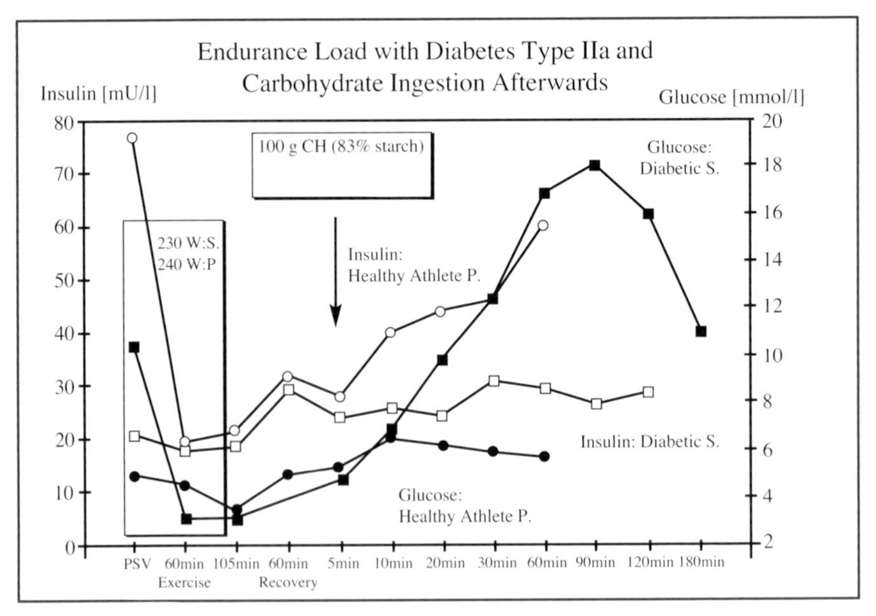

Fig. 30: Comparison of blood glucose and insulin in an endurance athlete (P.) and a type II diabetic who practises sport (S.). After 105 min endurance exersice, at lactate 2 mmol/l, both were given a solution with 100 g of complex carbohydrates to drink. The diabetic's blood glucose concentration rose sharply and the insulin did not rise. In contrast, athlete P.'s normal regulation showed a sharp rise in insulin and a low increase in blood glucose.

Athletes with diabetes II are completely capable of performance if they keep to their familiar dietary prescriptions and can (almost) go without medication if they carry out regular endurance training. **Self-measurement of the blood glucose concentration** (paper strip method) increases safety in cases of uncertain metabolic states before, during and after load. If a diabetic engages in a longer lasting endurance training session he must **protect** himself against **hypoglycaemia** by always having glucose handy.

12 Optimum Body Weight

As many people have discovered for themselves, optimum body weight is not dependent on tables of ideal weight, but is decisively influenced by **one's physique or build**. A person's build is mainly influenced by their height and weight. The measuring unit for body mass is the kilogram. Colloquially body mass is referred to as body weight. In many sports **body mass** has a decisive influence on performance capacity, e.g. in combat sports or downward movements with sporting equipment. In order to judge more fairly sporting performance in relation to body mass, the **weight class sports** were introduced e.g. wrestling, weight lifting, judo, boxing.

In **gymnastics, figure skating** or **rhythmical sports gymnastics**, male and female athletes with large body mass have less chance of accelerating their body mass around their body axis than lighter athletes. This experience makes it easier to select athletes early for technical-compository sports in which athletes of small build and low body mass predominate.

In other sports the measurements of certain extremities have a favourable influence on performance capacity, such as leg length for the high jump or arm length for gymnasts doing gymnastics on the pommel horse.

Because of the significance of body weight and height for sporting performance capacity, various indices have been calculated using these basic measurements. Well known are the **Quetelet Index (g/cm), the Rohrer Index (g/cm3), the Kaup Index (g/cm2) and the Broca Index**.

Of these indices the Broca Index has established itself because it provides information about excess weight. The **Broca Index** allows a comparative **rough estimate** of **physique constitution** and variations from normal weight. The Broca Index is calculated as follows:

$$\textbf{Broca Index} = \frac{\textbf{Body mass (kg)}}{\textbf{Height in cm-100}}$$

With the **Broca Index normal and excess weight are estimated**. A person is considered overweight if they weigh more than their height in centimetres minus 100. 80 kg and 170 cm equals a Broca Index of 1.14 [80: (170-100)=1.14].

Put simply, 100 can be subtracted from the height in centimetres and the result considered a weight estimate (170 cm-100=70 kg).

By determining the Broca Index it is proved that when a value of 1.2 in men and 1.3 in women is exceeded, the **health risk increases**, and that with above average frequency metabolism illnesses (Diabetes mellitus, raised blood fat, gout) increase.

Increasingly in scientific papers the **Body Mass Index (BMI)** is being used as the unit for measurement for **estimation of risk factors**. The BMI has proved more reliable and accurate than all other indices for estimating weight variations among adults.

The Body Mass Index is the quotient of body mass in kg/height in metres squared.

$$\textbf{Body Mass Index (BMI)} = \frac{\text{Body mass (kg)}}{\text{Height (m}^2\text{)}} = \frac{80\ \text{kg}}{1.90\ \text{m} \times 1.90\ \text{m}} = \mathbf{22.2}\ \textbf{(normal value)}$$

$$\textbf{BMI} = \frac{95\ \text{kg}}{1.90\ \text{m}^2} = \mathbf{26.3}\ \textbf{(overweight)}$$

The **normal values** of the BMI for women are considered to be 22 to 22.5 kg/m^2 and for men 24 to 24.5 kg/m^2. Values of over 23 for women and over 25 for men are already hints of a beginning overweight state. At a BMI of 25 kg/m^2 the insulin sensitivity of the muscle cells worsens and the risk of getting Diabetes mellitus type II increases considerably. At a BMI of over 30, medically prescribed and **controlled weight reduction** is necessary because the risk factors mentioned above increase by 30 to 60%. In special clinics body mass reduction of 20 to 30 kg in three months is possible. This weight reduction is achieved using **formula diets** where only 400 kcal daily are ingested under clinical conditions. This low energy ingestion consists of 50 g of proteins, 50 g of carbohydrates and 10 g of essential fatty acids. These clinical weight reduction programmes are patented as OPTI-FAST programmes.

Another estimation unit for excess weight **is the Waist-Hip ratio**. The waist measurement should not be greater than the hip measurement (measurement of abdomen and hips in cm). Women should not exceed a value of 0.8 and men of 1.0 if they want to be considered still of normal weight. Ideal values for women are < 0.8 and for men < 0.9. In sport the nutritional state can also be assessed by determining body fat mass. Several methods are suitable for this. Certain readings are provided by using a **calliper** to directly measure skin fold thickness in several places around the body. The measuring points on the body are laid down; examiners can choose between four to twelve measuring points. This procedure has the advantage over other methods of measuring body fat (impedance measurement, body density measurement, sonography) that the thickness of the subcutaneous fat layer over the particular part of the body can be measured directly.

Sport practitioners have average **total body fat** of 8-15 kg, whereby in endurance sports the lower values predominate. Female gymnasts as well as long distance runners are well below this at 3-5 kg. Weight lifters (without upper weight classes) have up to 20 kg of body fat.

Obesity **(adipositas)** begins when in men the proportion of body fat exceeds 23% and when women exceed 31%. According to current findings, in a certain frequency of cases adipositas has an **hereditary component**; with this inherent obesity normal slimming efforts have no effect. The substance leptin has entered the focus of scientific discussion. It is known from animal experiments that the **repletion factor leptin** can influence both fat and carbohydrate metabolism. A defect of the leptin receptor in the hypothalamus leads to disturbed feedback when satiety is reached.

Another new principle in treating overweight people has been made possible by the substance **silbutramin** (serotonin-noradrenaline reabsorption impeder). Silbutramin (Reductil(r)) increases the feeling of satiety and simultaneously increases energy utilisation on central regulation levels.

Meanwhile, because of the lifestyle and lack of exercise, 18% of all Germans are overweight and their treatment causes 5.4% of the costs of the health system, or considerably more than 10 billion DM per year. Reducing the number of overweight persons by only 1% could help save the German health system 825 million DM/year. According to analyses of the Federal Ministry of Health (1990) 1/3 of all costs to the health system are caused by illnesses related to diet.

The numerous **tables on normal weight** contain an error, they inaccurately take physique into consideration and in one calculatory variation they give preference to weight and in the other to height. Tables for ascertaining optimum weight have been calculated taking physique (build or stature) into consideration (MÖHR, 1988). The **physique type** can be differentiated into five structures (Fig. 31):

Physique Type I: Very slim, very little active body substance (muscle)
Physique Type II: Relatively slim, little active muscle mass
Physique Type III: Medium stature, balanced muscle mass
Physique Type IV: Relatively stocky, great deal of active muscle mass
Physique Type V: Very stocky (also athletic), very much active muscle mass

This method allows for fluctuations of 10% and differs from the selective way of looking at optimum weight as well as schematic height-weight prescriptions.

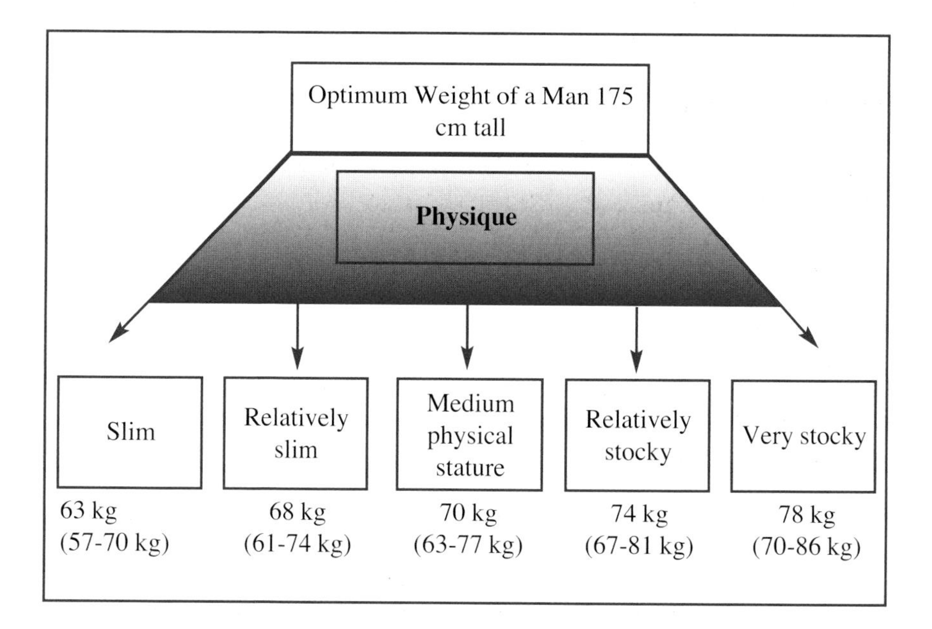

Fig. 31: Relationship between optimum weight and physique. Depending on body stature, with the same height the optimum weight can vary by an average of 15 kg.

13 Bibliography

ABRAMOVICZ, M.: Dehydroepiandrosterone (DHEA). In: The Medical Letter on Drugs and Therapeutics 38 (1996) 91-92.

ADNER, M./CASTELLI, W.P.: Elevated High-density Lipoprotein Level in Marathon-runners. In: J.Am. Med. Ass. 243 (1980) 534-536.

ADAMS GR./HADDAD F./BALDWIN KM.: Interaction of Chronic Creatine Depletion and Muscle Unloading: Effects on Postural Locomotor Muscles. In: J.Appl.Physiol. 77 (1994) 1198-1205.

ALBINA, J.E./MILLS, C.D./BARBUL, A./THIRKILL, C.E./HENRY, W.L./MASROFRANCESCO, B.: Arginine Metabolism in Wounds. In: Am. J. Physiol. Endocrinol. Metab. 245 (1988) E459-E467.

ANDERSON, R.A.: New Insights on the Trace Elements Chromium, Copper and Zinc, and Exercise. In: BROUNS, F., SARIS, W.H.M., NEWSHOLME, E.A. (eds.). Advances in Nutrition and Top Sport. In: Med. Sport Sci. 32 pp. 38-58 Karger, Basel 1991.

APFELBAUM, M.: "Wer keine angeborene Cholesterinkrankheit hat, sollte sich nicht um seinen Cholesterinspiegel kümmern". In: Jatros Kardiologie 3 (1994) Supplement 10-12.

ARENAS, J./RICOY J.R./ENCINAS, A.R../POLA, P./D'IDDIO, S./ZEVIANI, M./DIDONATO, S./ CORSI, M.: Carnitine in Muscle, Serum, and Urine of Nonprofessional Athletes: Effects of Physical Exercise, Training, and L-Carnitine Administration. In: Muscle & Nerve 14 (1991) 598-604.

ARMSEY, T.D./GREEN, G.A.: Nutrition Supplements. In: The Physician and Sportsmedicine. 25 (1997) 77-92.

ARNDT, K.: Leistungssteigerung durch Aminosäuren. 6.Aufl. Novagenics Verl., Arnsberg 1994.

ASSMANN, G.: Fettstoffwechselstörungen und koronare Herzkrankheit. 2.Aufl. MMV Medizin Verlag, München 1991.

BALSOM, P.D./EKBLOM, B./SÖDERLUND, K./SJÖDIN, B./HULTMAN, E.: Creatine Supplementation and Dynamic High-intensity Intermittent Exercise. In: Scand. J. Med. Sci. Sports 3 (1993) 143-149.

BALSOM, P.D./SÖDERLUND, K./EKBLOM, B.: Creatine in Humans With Special Reference to Creatine Supplementation. In: Sports med. 18 (1994) 268-280.

BÄSSLER, K.-H.: Nicht immer ist der Schlankheitswahn schuld. In: Therapiewoche 42 (1992) 2182-2190.

BÄSSLER, K.-H./GRÜHN, E./LOEW, D./PIETRZIK, K.: Vitamin-Lexikon. G.Fischer Verlag, Stuttgart-Jena-New York 1992.

BARBUL, A.: Arginine: Biochemistry, Physiology, and Therapeutic Implications. In: J. Parenteral Enteral Nutr. 10 (1985) 227-238.

BEEK, E.J. VAN DER: Vitamin Supplementation and Physical Exercise Performance. In: J.Sports Sci. 9 (1991) 77-89.

BERG, A./SIMON-SCHNAß, I./ROKITZKI, L./KEUL, J.: Die Bedeutung des Vitamin E für den Sportler. In: Dtsch. Z. Sportmed. 38 (1987) 416-422.

BERG, A./KÖNIG, D./HALLE, M./GRATHWOHL, D./BERG, AN./WEINSTOCK, C./NORTOFF, H./KEUL, J.: Wirkung eines biologischen Kombinationspräparates auf Enzym-Hefezellbasis auf Muskelstress und Immunsystem. In: Deutsche Z. Sportmedizin 48, (1997) 433-441.

BERGER, M.: Das Cholesterin – in einem anderem Licht gesehen. In: Jatros Kardiologie 3 (1994) Supplement 5-6.

BERGHOLD, F./PALLASMAN, K.: Aspekte der Höhenanpassung und der akuten Adaptationsstörung beim Bergsport in extermen Höhenlagen. In: Dtsch. Z. Sportmed. 34 (1983) 237-244.

BERGSTRÖM, I./HERMANSEN, L./HULTMAN, E./SALTIN, B.: Diet, Muscle Glycogen and Physical Performance. In: Acta Physiol. Scand.71 (1967) 140-150.

BESSMAN SP./SAVABI F. (1988): The Role of the Phosphocreatine Energy Shuttle in Exercise and Muscle Hypertrophy. In: TAYLOW AW/GOLLNICK PD/GREEN HJ/IANUZZO CD/NOBLE EG/METIVIER G/ SUUTON JR (eds): International Series on Sports Sciences. Vol. 21, pp 109-120. Human Kinetics, Champaign.

BEUKER, F.: Veränderungen der Haut und Hautanhangsorgane durch Missbrauch anaboler Steroide bei Sportlern. In: Haut 5 (1992) 6-15.

BIGARD, A.X./SATABIN, P./LAVIER, P./CANNON, F./TAILLANDER, D./GUEZENNEC, C.Y.: Effects of Protein Suplementation During Prolonged Exercise at Moderate Altitude on Performance and Plasma Amino Acid Pattern. In: Eur. J.Appl.Physiol. 66 (1993) 5-10.

BLOCK, K.P./BUSE, M.G.: Glucocorticoid Regulation of Muscle Branched-chain Amino Acid Metabolism. In: Med. Sci. Sports Exerc. 22 (1990) 316-324.

BLOMSTRAND, E./HASSMEN, P./ECKBLOM, B./NEWSHOLME, E.A.: Administration of Aamino Acids during Sustained Exercise-effects on Performance and on Plasma Concentration of Some Amino Acids. In: Eur. J. Appl. Physiol. 63 (1991) 83-88.

BÖHMER, D.: Der Einfluss des Hochleistungstrainings auf den Wasser-Salz-Haushalt. In: RIECKERT, H.(Hrsg.): Sport an der Grenze menschlicher Leistungsfähgkeit. Springer, Berlin-Heidelberg-New York 1981.

BRAUN, B./CLARKSON, P./FREEDSON, P. ET AL.: The Effect of Coenzym Q10 Supplementation on Exercise Performance. VO_2 max, and Lipid Peroxidation in Trained Cyclists. In: Int. J. Sport Nutr. 1 (1991) 353-365.

BROUNS, F./SARIS, W.H.M./NEWSHOLME, E.A. (EDS.): Advances in Nutrition and Top Sport. Karger, Basel (Med. Sport Sci.) 1991.

BROUNS, F.: Die Ernährungsbedürfnisse von Sportlern. Springer Verlag, Berlin 1993.

BROUNS, F./KOVACS, E.: Rehydratationsgetränke für Sportler. In: TW Sport + Medizin 8, (1996) 167-174.

BUCCI, L./HICKSON, J.F./PIVARNIK, J.M./WOLINSKY, J.C./MCMAHON, J.C./TURNER, S.D.: Ornithine Ingestion and Growth Hormone Release in Bodybuilders. In: Nutrition Res. 10 (1990) 239-245.

BURKE, L.M./READ, S.D.: Dietary Supplements in Sport. In: Sports Medicine 15 (1993) 43-65.

BURKE L.M. ET AL.: Muscle Glycogen Storage after Prolonged Exercise: Effect of the Glycämic Index of Carbohydrate Feedings. In: J. Appl. Physiol. 75 (1993) 1019-1023.

CAMPBELL, W.W./ANDERSON, R.A.: Effects of Aerobic Exercise and Training on the Trace Minerals, Chromium, Zinc and Copper. In: Sports Med. 4 (1987) 9-18.

CASEY, A./CONSTANTIN-TEODOSIU, D./HOWELL, S./HULTMAN, E./GREENHAFF, P.L.: Creatine Supplementation Favourably Affects Performance and Muscle Metabolis during Maximal Intensity Exercise in Humans. In: Am. J. Physiol. 271 (1996) E31-E37.

CLARKS, N.: Sports Nutrition Guide book. Human Kinetics Publishers. Champaign 1990.

CLAUSTRAT, B./BRUN, J./CHAZOT, G.: "Melatonin and Jet Lag": Confirmatory Result Using a Simplified Protocol. In: Biological Psychiatry 32 (1992) 705-711.

COGGAN, A.R../SWANSON, S.C.: Nutritional Manipulation before and during Endurance Exercise: Effects on Performance. In: Med. Sci. Sports Exerc. 24 (1992) Supplement S331-S335.

COSTILL, D.L./COYLE, E.F./DALSKY, G./EVANS, W./FINK, W./HOOPES, D.: Effects of Elevated Plasma FFA and Insulin on Muscle Glycogen Usage during Exercise. In: J. Appl.Physiol. 43 (1977) 695-699.

COYLE, E.F./HAGBERG, J.M./HURLEY, B.H./MARTIN,III, W.H./EHSANI, A.A./HOLLOSZY, J.O.: Carbohydrate Feeding during Prolonged Strenous Exercise Can Delay Fatigue. In: J. Appl. Physiol. 15 (1983) 466-471.

DEGKWITZ, E.: Neue Aspekte der Biochemie des Vitamin C. Z. In: Ernährungswiss. 24 (1985) 219-230.

DELANGHE, J./DE SLYPERE, J.P./DE BUYZERE, M. ET AL.: Normal Reference Values for Creatine, Creatinine and Carnitine Are Lower in Vegetarians. In: Clin. Chem. 35 (1989) 1802-1803.

DENNING, H.: Über Steigerung der körperlichen Leistungsfähigkeit durch Eingriffe in den Säurenbasenhaushalt. In: Medizinische Wochenschrift 19 (1937) 733-736.

DIEBSCHLAG, W.: Die optimale Ernährung für Sportler. In: Leistungssport 1 (1985) 15-22.

DONATH, R./SCHÜLER, K.-P.: Ernährung der Sportler. 2.Aufl. Sportverl., Berlin 1979.

DONIKE, M./RAUTH, S.: Dopingkontrollen. Bundesinstitut Sportwissenschaft, Köln 1992.

DRAGAN, G.I./VASILIU, A./GEORGESCU, E./DUMAS, I.: Studies Concerning Chronic and Acute Effects of L-Carnitine on Some Biological Parameters in Elite Athletes. In: Physiologie 24 (1987) 23-28.

EISENMAN, P.A./JOHNSON, S.C./BENSON, J.E.: Coaches Guide to Nutrition and Weight Control. Leiser Press. Champaign 1990.

EISINGER, M.: Vergleichende Untersuchung der Ernährungszufuhr zweier Kostformen (konventionelle Sportkost und ovo-lacto-vegetarische Vollwertkost) bei einem Ultralangstreckenlauf (Deutschlandlauf 1987). Ernährungswissenschaftliche Schriftenreihe. Wissenschaftlicher Fachverlag, Gießen 1990.

EISINGER, M./LEITZMAN, C.: Ernährung und Sport – eine Übersicht. Dtsch.Z. Sportmed. 43 (1992) 472-493.

ENGELHARDT, M./NEUMANN, G.: Sportmedizin. Grundlagen für alle Sportarten. BLV-Verlag, München 1994.

FLECK, S.J./REIMERS, K.J.: The Practice of Making Weight:Does It Affect Performance? In: Strength and Conditioning (Colorado Springs) 16 (1994) 66-67.

FÖHRENBACH, R./MÄRZ, W./LOHRER, H./SIEKMEIER, R./EVANGELIOU, A./BÖHLES, H.: Der Einfluß von L-Carnitin auf den Lipidstoffwechsel von Hochleistungssportlern. Dtsch. Z. Sportmedizin 44 (1993) 349-356.

FRIEDRICH, W.: Handbuch der Vitamine. Urban u. Schwarzenberg, München 1987.

FRÖHNER, G.: Die Belastbarkeit als zentrale Größe im Nachwuchstraining. Trainerbibliothek 30, Philippka, Münster 1993.

GEIß, K.-R./HAMM, M.: Handbuch der Sportler-Ernährung. Behrs Verlag, Hamburg 1990.

GRAHAM, T.E./SPRIET, L.L.: Performance and Metabolic Responses to a High Caffeine Dose during Prolonged Exercise. In: J. Appl. Physiol. 71(1991) 2292-2298.

GREEN, A. L./SIMPSON, E. J./LITTLEWOOD, J. J./MACDONALD, I. A./GRENHAFF, P.L.: Carboydrate Ingestion Augments Creatine Retention during Creatine Feeding in Man. In: Acta Physiol. Scand. 158 (1996) 195-202.

GREENHAFF, P.L./CASEY, A./SHORT, A.H./HARRIS, R./SÖDERLUND, K./HULTMAN, E. : Influence of Oral Creatine Supplementation of Muscle Torque during Repeated Bouts of Maximal Voluntary Exercise in Man. In: Clin. Sci. 84 (1993) 565-571.

GREENHAFF, P.L./BODIN, K./SÖDERLUND, K./HULTMAN, E.: Effect of Oral Creatine Supplementation on Skeletal Muscle Phosphocreatine Resynthesis. In: Am. J. Physiol. 266 (1994) E725-E730.

GREENHAFF, P./CASEY, A./GREEN, A.: Creatine Supplementation Revisited: A Update. In: Insider Special 4 (1996) 1-2.

GRUBE,W.: Der Kaliumgehalt von Läufern. In: Ausdauersport,H .5. Schriftenreihe des Verbandes langlaufender Ärzte. S. 35-56, Presse-Druck, Augsburg 1993.

HAAS, R.: Dr. Haas Top Diät. BLV Verlagsgesellschaft, München 1986.

HAAS, R.: Die Dr. Haas Leistungsdiät. BLV Verlagsgesellschaft, München 1986.

HALLMARK, M.A./REYNOLDS, T.H./DESOUZA, CH.A./DOTSON, CH.O./ANDERSON, A./ ROGERS, M.A.: Effect of Chromium and Restriktive Training on Muscle Strength and Body Composition. In: Med. Sci. Sports Exerc. 28, (1996) 139-144.

HAMM, M.: Ernährung des (Hoch)Leistungssportlers in der Trainings- und Wettkampfphase. In: Akt. Ernährungs.-Med. 16 (1991) 73-77.

HAMM, M./WEBER, M.: Sporternährung. 2.Aufl. Walter Hädecke Verlag, Weil der Stadt 1988.

HARGREAVES, M./COSTILL, D.L./COGGAN, A.R./FINK, W.J./NISHIBATA, I.: Effects of Carbohydrate Feedings on Muscle Glycogen Utilization and Exercise Performance. In: Med. Sci. Sports Exerc. 16 (1984) 219-222.

HARGREAVES, M./COSTILL, D.L./FINK, W.J./KING, D.S./FIELDING, R.A.: Effect of Pre-exercise/ Arbohydrate Feedings on Endurance Cycling Performance. In: Med. Sci. Sports Exerc. 19 (1987) 33-36.

HARRIS, C./SÖDERLUND, K./HULTMAN, E.: Elevation of Creatine in Resting and Exercising Muscle of Normal Subjects by Creatine Supplementation. In: Clinical Science 83 (1992) 367-374 .

HARRIS, R. C./HULTMAN, E./NORDESJÖ, L. O.: Glycogen,.Glycogenolytic Intermediates and High Energy Phospates Determined in Biopsy Samples of Musculus Quadriceps Femoris. In: Scand. J. Clin. Lab. Invest. 33 (1974)109-120.

HELFER, S.K./WIDEMAN, L./GAESSER, G.A./WELTMAN, A.: Branched-chain Amino Acid (BCCA) Supplementation Improves Endurance Performance in Competitive Cyclists. In: Med. Sci. Sports Exerc. 27 81995) Suppl. S149.

HERTOG, M.G.L./FRESKENS, E.J.M./HOLMAN, P.C.H./KATAN, M.B./KROMHOUT, D.: Dietary antioxidant Flavonoids and Risk of Coronary Heart Disease: The Zutphen Elderly Study. Lipids 342(1993) 1007-1011.

HOFFMANN, D.: Der Körpereigenstatus bei Sporttreibenden und seine Beziehung zur körperlichen Belastungs- und Leistungsfähigkeit. Verlag Shaker, Frankfurt 1995.

HOLTMEIER, H.I.: Gesunde Ernährung. 3. Aufl. Springer Verlag, Berlin 1995.

HULTMAN, E.: Diatary Manipulation an Aid to Preparation for Competitions. In: Proceedings of the World Conference on Sports Medicine. pp. 239-265, Melbourne 1974.

HULTMAN, E./SODERLUND K./TIMMONS, K./CEDERBLAD, G./GREENHAFF, P.L.: Muscle Creatine Loading in Man. In: J. Appl. Physiol. 81 (1996) 232-237.

HUTH, K./KLUTHE, R.(HRSG.): Lehrbuch der Ernährungstherapie. Thieme, Stuttgart 1986.

IVY, J.L./MILLER, W./DOVER, V./GOODYEAR, L.G./SHERMAN, W.H./WILLIAMS, H.: Endurance Improved by Ingestion of a Glucose Polymer Supplement. In: Med. Sci. Sports Exerc. 15 (1983) 466-471.

JUNG, K.: Sport und Ernährung. Meyer und Meyer, Aachen 1984.

KARLSSON, I.: Antioxidants and Exercise. Human Kinetics, Champaign 1997.

KEUL, J./WITZIGMANN, E.: Die Olympia-Diät. W. Heyne-Verlag, München 1988.

KOLLER, A./MAIR, J./JUDMAIER, W. ET AL.: Der belastungsinduzierte Muskelschaden - Neue Wege in der Diagnostik und der Lokalisation. In: Dtsch. Z. Sportmed. 45 (1994) 346-358.

KONOPKA, P.: Sporternährung. 5.Aufl. BLV Sportwissen, München 1994.

LEIBOVITZ, B.E.: L-Carnitine. Lonza, Basel 1993

LEVINE, L./EVANS, W.J./CADARETTE, B.S./FISHER, E.C./BULLEN, B.A.: Fructose and Glucose Ingestion and Muscle Glycogen Use during Submaximal Exercise. In: J. Appl. Physiol. 55 (1983) 1767-1771.

LINDEMAN, A.K.: Nutritient Intake of an Ultradurance Cyclist. In: Int. J. Sport Nutrition 1 (1991) 79-85.

LINO, A./SILVY, S./RUSCONI, A.: "Melatonin and Jet Lag: Treatment Schedule". In: Biological Psychiatry 34 (1993) 587-588.

LITHELL, H./CEDERMARK, M./FRÖBERG, I./TESCH, P./KARLSSON, I.: Increase of Lipoprotein-lipase Activity in Skeletal Muscle during Heavy Exercise. Relation to Epinephrine Excretion. In: Metabolism 30 (1981) 1130-1135.

MACLAREN, D./ADAMSON, G.: An Investigation into the Combined Effects of Creatine and Sodium Bicarbonate Supplementation on Repeated Bouts of High-intensity Exercise in Elite Cyclists. In: J. Sports Sci. 13 (1995) 429-432.

MAUGHAN, R.J.: Creatine Supplementation and Exercise Performance. In: Int. J. Sport Nutr. 5 (1995) 94-101.

MARCONI, C./SASSI, G./CARPINELLI, A./CERRETELLI, P.: Effects of L-Carnitine Loading on the Aerobic and Anaerobic Performance of Endurance Athletes. In: Eur. J. Appl. Physiol. 54 (1985) 131-135.

MARNIEMI, J./PELTONEN, P./VUORI,T./HIETANEN, E.: Lipoprotein Lipase of Human Postheparin Plasma and Adipose Tissue in Relation to Physical Training. In: Acta Physiol. Scand. 110 (1980) 131-135.

MITCHEL, M.E.: Carnitine Metabolism in Human Subjects. In: Am. J. Clin.Nutr. 31 (1978) 293-306.

MÖHR, M.: Körpergewicht Erwachsener (Optimalgewicht).In: DITTMER, A./ARKONA, S.(Hrsg.): Ärztetaschenbuch. 2.Aufl. Verl.Volk und Gesundheit, Berlin 1988.

NEUMANN, G./KÖHLER, E./KÄMPFE, U.: Gewichtsabnahme durch Formuladiät und Sport? In: TW Sport + Medizin 3 (1991) 444-447.

NEUMANN, G./PÖHLANDT, R.: Einfluss von Kohlenhydratgaben während Ergometerausdauerleistung auf die Fahrzeit. In: Schriftenreihe zur Angewandten Trainingswissenschaft. IAT Leipzig. 1 (1994) 7-26.

NEUMANN, G./SCHÜLER, K.-P.: Sportmedizinische Funktionsdiagnostik. J. A. Barth Verlag, Leipzig 1994.

NEWSHOLME, E.A./LEECH, A.R.: Biochemistry for Medical Science. Wiley, Chichester 1983.

NOAKES, T.: Lore of Running. 3.Auflage. Oxford University Press, Oxford 1992.

NOAKES, T.: Fluid Replacement during Exercise. In: Exercise and Sport Sciences Reviews. Holloszy, J.O. (ed.) Vol.21 (1993) 297-330. Williams & Wilkins, Baltimore 1993.

NOTHACKER, S.M.: Besonderheiten in der Ernährung des Leistungssportlers in der Trainings- und Wettkampfphase. In: Ernähr.-Umschau 39 (1992) 113-116.

OSHIMA, Y./MIYAMOTO, T./TANAKA, S./WADAZUMI, T./KURIHARA, N./FUJIMOTO, S.: Relationship between Isicapnic Buffering and Maximal Aerobic Capacity in Athletes. In: Eur. J. Appl. Physiol. 76 (1997) 409-414.

PARRY-BILLINGS, M./MACLAREN, D.P.M.: The Effect of Sodiumbicarbonate and Sodium Citrate Ingestion on Anaerobic Power during Intermittent Exercise. In: Eur.J.Appl. Physiol. 55 (1986) 524-529.

PARRY-BILLINGS, M./BUDGETT, R./KOUTEDAKIS, Y. ET AL.: Plasma Amino Acid Concentrations in the Overtraining Syndrome: Possible Effects on the Immune System. In: Med. Sci. Sports Exerc. 24 (1992) 1353-1358.

PAFFENBARGER, JR. R. S.: Die Rolle der körperlichen Aktivität in der primären und sekundären Prävention der koronaren Herzkrankheit. In: WEIDEMANN,H./SAMEK,L.(Hrsg.): Bewegungstherapie in der Kardiologie. Darmstadt, Steinkopf 1982.

PETERSON, M.K./PETERSON, O.: Eat to Compete. In: Year Book Medical Publishers, Chicago 1988.

PIENDL, A./RUMMEL-PITLIK, F.: Hefe – ein vielseitiges Lebens- und Heilmittel. Apotheker J. 6 (1985) 66-71.

REHRER, N.J./BROUNS, F./BECKERS, E.J. ET AL.: Gastric Emptying with Repeated Drinking during Running and Cycling. In: Int. J. Sports Med. 11 (1990) 238-243.

REITER, J.R./ROBINSON, J.: Melatonin. Knaur, München 1997.

REUSS, F.: Elektrolyt- und Flüssigkeitssubstitution beim Sportler in der Trainings- und Wettkampfphase. In: Ernähr.- Umschau 39 (1992) 117-S122.

ROKITZKI, L./KEUL, J.: Vitaminbedarf und Vitaminversorgung – im Sport unbekannte Größe? Dtsch.Z.Sportmed. 43 (1992) 524-527.

ROKITZKI, L.: Kalzium – nicht nur ein potentes Osteoporose-Antidot. In: TW Sport + Medizin 6 (1994) 53-56.

ROKITZKI, L./SAGREDOS, A.N./KECK, E./SAUER, B./KEUL, J.: Assessment of Vitamin B_2 Status in Performance Athletes of Varios Types of Sports. In: J. Nutr. Sci.Vitaminol. 40 (1994) 11-22.

ROKITZKI, L./ANDREE, N./SAGREDOS, N./REUß, F./BÜCHNER, M./KEUL, J.: Acute Changes in Vitamin B_6 Status in Endurance Athletes before and after a Marathon. In: Int. J. Sports Nutr. 4 (1994) 154-165.

ROCKITZKI, L./SAGREDOS, A.N./REUß, F./CUFFI, D./KEUL, J.: Assessment of Vitamin B_6 Status of Strength and Speedpower Athletes. In: J. Am. Coll. Nutrition 13 (1994) 87-94.

RÖCKER, K./OTTO, B./MAYER, F./STEHLE, P./DICKHUT, H.H.: Die Bedeutung der Nährstoffrelation bei Ausdauersportlerinnen. In: Ernähr.-Umschau 39 (1992) 109-112.

SAITO, H./TROCKI, O./WANG, S.: Metabolic and Immune Effects of Dietary Arginine Supplementation after Burn. In: Arch. Surg. 122 (1987) 784-789.

SARIS, W.H.M./VAN ERP-BAART/M.A., BROUNS, F./WESTERTERP, K.R./TEN HOOR, F.: Study on Food Uptake and Energy Expenditure during Extreme Sustained Exercise: the Tour de France.In: Int. J. Sports Med. 10 (1989) 25-S31.

SCHENA, F./GUERRINI, F./TREGNAGHI, P./KAYSER, B.: Branched-chain Amino Acids Supplementation during Trekking at High Altitude. In: Eur. J. Physiol. Occup. Physiol. 65 (1992) 394-398.

SCHRANZER, G.M.: Selen. 3.Aufl., J.A.Barth, Leipzig 1997.

SCHUSTER, H.G./NEUMANN, G./BUHL, H.: Kreatinin- und Kreatinveränderungen im Blut bei körperlicher Belastung. In: Med. u.Sport 19 (1979 235-240.

TANNENBAUM, S.R.: Preventive Action of Vitamin C on Nitrosamin Formation. In: Elevated Dosages of Vitamins. WALTHER, P./STÄHELIN, H./BRUBACHER, G. (eds.). Int. J. Vit. Nutr. Res., Suppl. 30 (1989) 109-113.

WEBSTER, J.M./SCHEETT, P.T./DOYLE, M.R./BRANZ, M.: The Effect of a Thiamin Derivative on Exercise Performance. In: Eur. J. Appl. Physiol. 75 (1997) 520-524.

WESTON, S.B./ZHON, S./WEATHERBY, R.P./ROBSON, S.J.: Does Exogenous Coenzyme Q_{10} affect Aerobic Capacity in Endurance Athletes? In: Int. J. Sport Nutr. 7(1997) 197-206.

WILLMORE, J.H./COSTILL, D.L.: Training for Sport and Activity. 3.Aufl. ,Brown Publ., Dubuque 1988.

WILLMORE, J.H./COSTILL, D.L.: Physiology of Sport and Exercise Champaign: Human Kinetics 1994.

WOLFRAM, G.: Vollwerternährung, vollwertige Ernährung. In: Akt. Ernährung 13 (1988) 43-46.

WORM, N.: Die Top-Sport-Diät für alle. Mary-Hahn-Verlag, München 1988.

WORM, N.: Ratgeber Ernährung. TR Verlagsunion. 2.Aufl., München 1992.

WORM, N./SCHRÖDER, E.-M.: Die Ausdauer-Vollwerternährung. Sportinform Verlag F. Wöllzenmüller, Oberhaching 1987.

WYNDHAM, C.H./STRYDOM, N.B.: Körperliche Arbeit bei hoher Temperatur. In: HOLLMANN, W. (Hrsg.):. Zentrale Themen der Sportmedizin. 3.Aufl. Springer, Berlin-Heidelberg-New York 1986.

ZIEGLER, R.: Selen – vom Insider-Tip zur Präventiv-Empfehlung. In: TW Sport + Medizin 9 (1997) 137-140.

14 Index

15 Abbreviations

AcetylCoA	"Activated acetic acid"
ADH	Antidiuretic hormone (hormone for regulating the water balance)
ADP	Adenosine diphosphate
ANP	Atrial natriuretic peptide (Hormone for regulating the water balance)
ATP	Adenosine triphosphate
BU	Bread Unit (12 g of carbohydrates)
BM	Body Mass
BMI	Body Mass Index
Ca	Calcium
CH	Carbohydrates
CHU (CU)	Carbohydrate Unit (10 g of carbohydrates)
CK	Creatine kinase
CNS	Central Nervous System
CoA	Coenzyme A
CP	Creatine phosphate
DGE	German Association for Nutrition (Deutsche Gesellschaft für Ernährung)
DNA	Deoxyribonucleic acid
2.3-DPG	2.3-Diphosphoglycerate
Fe	Iron
FFA	Free Fatty Acids
FTF	Fast twitch muscle fibres
GI	Glycaemic Index
GRH	Growth hormone
IOC	International Olympic Committee
K	Potassium
Mg	Magnesium
Na	Sodium
NAD	Nicotinic acid amide adenine dinucleotide
NADP	Nicotinic acid amide adenine dinucleotide phosphate
P	Inorganic phosphate
STF	Slow twitch muscle fibres
TG	Triglycerides (neutral fats)
WHO	World Health Organisation
Zn	Zinc

16 Glossary of Important Technical Terms

Basic Nutrients:
The three basic nutrients are: Carbohydrates, proteins and fats. The essential nutrients include certain amino acids, certain fatty acids, vitamins, minerals, trace elements and water. The nutrients aiding bodily functions are fibres, aromatics and flavourings, colourings and active substances.

Basic Utilisation:
Minimum energy utilisation to maintain normal ability to live is usually 1,500-1,700 kcal/day. Physical load increases performance utilisation, which is in relation to intensity and duration.

Body Mass Index:
Quotient from body mass in kg over height squared. Values below 23 normal, over 23-25 excess weight begins and over 30 obesity (adipositas).

Electrolytes:
Minerals which because of their electric charge change to anodes (anions) or cathodes (cations). Important electrolytes in sport are sodium, potassium, calcium and magnesium, which are also the main components of electrolyte solutions.

False Nutrition:
Deviation from normal (optimal) nutrient ingestion which leads to excess weight, performance restrictions or illnesses.

Formula Diet:
Industrially manufactured preparations for weight reduction. They contain all vital active substances (vitamins, minerals) with an energy quantity of 400-800 kcal/day.

Fructose:
Widespread monosaccharide in plants and honey. Component of household sugar (cane sugar). When ingested it must first be changed into glucose in the liver. Causes no rise in insulin when consumed. Ingestion of high fructose concentrations (over 3.5%) during load usually causes indigestion (diarrhoea).

Glucagon:
Peptide hormone produced in the A cells of the pancreas which stimulates the release of free fatty acids and the production of ketone bodies in the liver when blood sugar drops. Glucagon increases in the blood when insulin decreases.

Gluconeogenesis:
Production of glucose in liver and kidneys from amino acids, glycerol and lactate. Reverse metabolism direction of glycolysis.

Glucose:
Important monosaccharide in body tissues and blood (blood glucose). Next to fructose a component of household sugar. Glucose is the direct energy supplier in metabolism during muscle work. The storage form of glucose is glycogen.

Glycaemic Index (GI):
Effects of ingested carbohydrates on blood glucose concentration. Glucose ingestion causes a higher glycaemic index than starch ingestion.

Glycogen:
Storage form of glucose in tissue, especially in muscles and liver. Endurance training enlarges the glycogen stores.

Glycogen depletion:
Emptying (breakdown) of the glycogen stores in loaded muscles and liver during longer sporting performance.

Glycolysis:
Breakdown of glycogen or glucose without oxygen (anaerobic) to the stage of pyruvate or lactate.

Insulin:
Hormone of the B cells of the pancreas necessary for sweeping glucose out of the blood into the muscle cells and organs at rest. Insulin deficiency or insulin insensitivity of the tissue leads to diabetes (Diabetes mellitus).

"Jetlag":
State of great fatigue and reduction in performance after intercontinental flights; more pronounced after flights opposite to the course of the sun (eastwards). One day of adjustment is to be reckoned with for every hour of time difference.

Lactate (Salt of lactic acid):
Intermediate product of metabolism by breakdown of glucose without oxygen (anaerobic). Occurs during intensive muscle work and is used especially in competitive sport as an indicator for controlling load intensity in the training areas. Measurement parameter for the definition of metabolism thresholds.

Melatonin:
Hormone of the pineal gland responsible for maintaining the day-night rhythm. Sold in the USA as a dietary food. If taken on arrival it can ease adjustment to time differences after intercontinental flights.

Micronutrients:
Mineral and vitamin content in food.

Nutrient density:
Relationship of the vitamins and minerals contained in nutrients per 1,000 kcal. Low nutrient density is found e.g. in household sugar and high density in bananas or potatoes. Also seen as the desirable nutrient content per ingested calorie.

Nutrient ratio:
Proportions of carbohydrates, proteins and fats in food which cover energy requirements (energy percentage of total energy consumption/calorie requirements).

Osteoporosis:
Reduced bone mass in comparison to the norm with decrease in loadability (increased danger of breaking). Afflicts women more than men because of oestrogen deficiency and lack of exercise.

Risk factors:
Deviations from normal lifestyle due to false nutrition, lack of exercise, smoking and/or stress. Consequences are cardiovascular and metabolism illnesses with the risk of major blood supply problems to the heart (heart attack).

Supplementation:
Supplementation of natural diet with certain active substances, e.g. vitamins or minerals (trace elements).

Trace elements:
Inorganic substances (minerals) which are usually vital to life and are ingested in quantities less than 100 mg/day.

Vegetarianism:
Diet with emphasis on carbohydrates and ingestion of proteins of plant origin. From animals only milk and/or eggs are consumed. The extreme form of vegetarianism is the avoidance of milk and eggs and the exclusive ingestion of plant proteins (vegans).

Vitamins:
Essential active substances, the supply of which the body is dependent on because it needs them for functions vital to life. Vitamin deficiency leads to poorer performance and illnesses. High vitamin consumption does not improve sporting performance.

Weight Reduction:
Reduction of body mass (body weight) by fasting (totally or partially), ingestion of reduced energy mixed diet, diets with extreme nutrient ratios or industrially manufactured nutrient mixtures (formula diet), fruit and rice days as well as outsider diets. Dietary measures are more effective in combination with increased physical load.

Whole food nutrition:
Variations of vegetarian diet through deliberate consumption of unaltered foods, dressed foods, processed foods, temperature treated foods and/or finished products (ready to eat).